Basic Concepts in Organic Chemistry

Basic Concepts in Organic Chemistry

A programmed learning approach

Peter Simpson

University of Sussex
UK

CHAPMAN & HALL
University and Professional Division
London · Glasgow · Weinheim · New York · Tokyo · Melbourne · Madras

Published by Chapman & Hall, 2-6 Boundary Row, London SE1 8HN, UK

Chapman & Hall, 2-6 Boundary Row, London SE1 8HN, UK

Blackie Academic & Professional, Wester Cleddens Road, Bishopbriggs, Glasgow G64 2NZ, UK

Chapman & Hall GmbH, Pappelallee 3, 69469 Weinheim, Germany

Chapman & Hall USA., One Penn Plaza, 41st Floor, New York, NY10119, USA

Chapman & Hall Japan, ITP - Japan, Kyowa Building, 3F, 2-2-1 Hirakawacho, Chiyoda-ku, Tokyo 102, Japan

Chapman & Hall Australia, Thomas Nelson Australia, 102 Dodds Street, South Melbourne, Victoria 3205, Australia

Chapman & Hall India, R. Seshadri, 32 Second Main Road, CIT East, Madras 600 035, India

First edition 1994
Reprinted 1994

Typeset in Great Britain by ROM-Data Corporation Ltd, Falmouth, Cornwall
Printed in Great Britain by The Alden Press, Oxford

ISBN 0 412 55830 0

A Catalogue record for this book is available from the British Library

Library of Congress Cataloging-in-Publication Data
Simpson, Peter.
Basic concepts in organic chemistry:a programmed learning approach / Peter Simpson. - 1st ed.
p. cm.
Includes index.
ISBN 0-412-55830-0
1. Chemistty, Organic-Programmed instruction.I. Title.
QD253.S53 1994
547'.007'7-dc20 93-21314
CIP

Printed on permanent acid-free text paper, manufactured in accordance with ANSI/NISO Z39.48-1992 and ANSI/NISO Z39.48-1984 (Permanence of Paper).

Contents

Preface

A number of years ago, people wrote numerous learning programmes in chemistry. It became fashionable, and consequently many were written for the sake of writing programmes rather than to meet an identified need. The programmes faded away.

Since that time, few have been written. Dr Stuart Warren, having identified a need, was perhaps the first to reintroduce them. His programmed approaches to learning the mechanistic chemistry of the carbonyl group[1] and organic synthesis[2] are well known and well used. A few years ago I came across a programme he had written for his own students on the use of 'curly arrows' in organic reaction mechanisms. With this in mind, as well as some clearly identified needs, I developed our own 'curly arrow' programmes, as well as a number of others in appropriate areas of basic organic chemistry, for use here at Sussex. These programmes are presented here in the hope that students in other universities and colleges will find them as useful as do students at Sussex.

Several of the examples used by Dr Warren in his 'curly arrow' programme I borrowed for this purpose, and I am grateful to him for his permission to retain these examples in Programme 4 (frames 25, 30, 31, 32) of this published suite of programmes. I am also grateful, of course, to our own students who have played such an important role in putting the programmes to the test, and in improving earlier versions. My colleagues, with whom the programmes have been discussed, I also thank for all that I have learnt from them, and I thank Peter Knowles and Peter Simmons for their help and advice with both the hardware and the software used to prepare the diagrams. The use of Microsoft's ChemWindow in the preparation of these diagrams is also gratefully acknowledged.

I hope you find these programmes as useful as our own students have. Good luck with them, and with your learning of the whole art and science of organic chemistry!

Peter Simpson
Brighton

[1]Warren, S. (1974) *Chemistry of the Carbonyl Group: A Programmed Approach to Organic Reaction Mechanisms*, Wiley, Chichester.

[2]Warren, S. (1978) *Designing Organic Syntheses: A Programmed Introduction to the Synthon Approach*, Wiley, Chichester.

Introduction

Programmed learning in organic chemistry – some basic concepts

The concepts treated here are fundamental foundations for any degree course in organic chemistry; they must really be mastered early on if subsequent courses are to be found interesting and rewarding. Not all concepts are susceptible to this treatment, of course, and it is, in any case, good for students to vary their learning techniques. So, static stereochemistry is not treated here, for example, partly because it would demand a very long programme, and partly because it is better learnt by handling models, by means of problem classes and by the more traditional lecture/tutorial approach. However, in our experience, some of the more dynamic aspects of stereochemistry, the changes in stereochemistry accompanying reactions for example, are ideally suited to the programmed learning approach. The programmes, and the concepts covered therein, are listed separately, on the contents page, and a few explanatory comments on each are offered there.

Within some of these concepts certain intellectual techniques have been developed to enable them to be simply handled. Thus, for example, 'curly arrows' may be used to derive resonance forms for molecules, or to describe the progress (mechanism) of reactions. Often, however, these techniques are taught, as it were, on the hoof, the assumption being that students can easily pick them up as they go along. However, this assumption is very often wrong, and these programmes have been developed, here at Sussex, so that the techniques and the concepts can be properly addressed and digested.

Some students find both the techniques and the concepts easy to master, but many others find them difficult. The opportunity to work at your own pace, which the programmes offer, is invaluable, and we have found the programmes to be both popular with students and effective. The general response has been 'Why can we not have more of them?'.

Some people, therefore, may find the programmes (or some of them) easy; others may find them tougher. How hard, or how easy you find them will, of course, depend in part upon your background, and whether you have come across the ideas before. However, no matter how easy or how tough you find the programmes, they should help you to master the concepts and to develop confidence in your chemistry.

The programmes are built up from **frames**. Each frame tells you something about the concept, and usually includes a question to help you to find out whether or not you have understood the point. You should not proceed with the next frame until you have arrived at the correct solution to the question in the last one. If you cannot get that question right, then talk to your friends or to a tutor. Since there is no time limit on the exercise, this is quite easy to do. The opportunity to proceed at your own pace, with no one 'breathing down your neck', is one of the major advantages of programmes like these.

To use a programme, cover the first page with a sheet of paper and then pull down that sheet to reveal the first frame. Each frame ends with a row of asterisks, and this marks the position you should pull the paper down to. When you have arrived at an answer to the question which concludes the frame write it down on a separate sheet of paper (this is very important). Then pull down the paper to the next row of asterisks and check your answer against the one given at the beginning of the newly revealed frame. *Questions and answers are highlighted by being printed in italics.* In this manner, work your way through each programme, frame by frame, always obtaining and understanding the correct answer to one frame before passing on to the next.

Ideally, the programmes should be tackled alongside, or just before, your organic chemistry courses which introduce, or depend upon, the relevant concepts for the first time. Since the programmes are, largely, built one upon the other, it makes sense to do them in the order in which they appear here. Only if you are already master of the concepts and skills which it contains should you skip a programme. Programme 7 is probably less dependent upon its predecessors than are the others, but that is largely because it is dealing with rather different sorts of ideas.

It is also a good idea to have your organic textbook handy as you proceed through the programmes, so that you may consult it to clear up any point which you have not fully understood. Consultations with your fellow students may also be profitable, and, of course, you can always approach your tutor for advice.

Programme 1

Introduction to resonance and delocalization – Part I

For a given molecular structure, it is often possible to distribute any π-bonding electrons or lone pairs in more than one way, whilst maintaining the integrity of the structure. It is usually possible to distribute the σ-bonding electrons in only one way, so that we can presently confine our attention to lone pairs and π-electrons.

For example, we could write acetone (propanone) as:

1

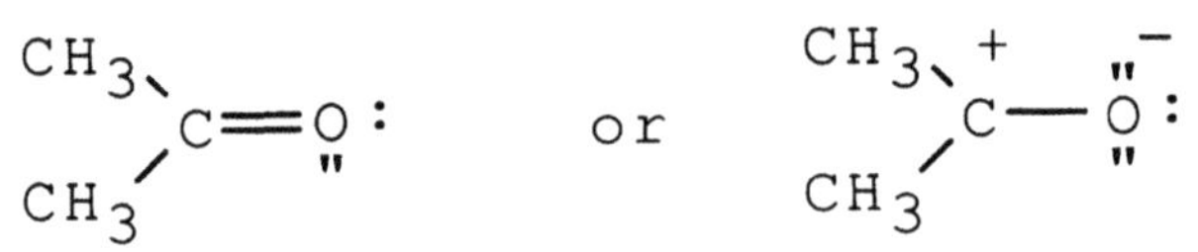

A B

We have shown the lone pairs (of electrons) here, although this is not standard practice. Since this makes it easier to see how to write resonance forms we shall continue to do so throughout Part I but, in Part II, when the basic ideas have become familiar, we shall drop the practice.

? *There is one other way, form C, in which we might have written the acetone molecule. What is it?*

* *

You should have written:

2

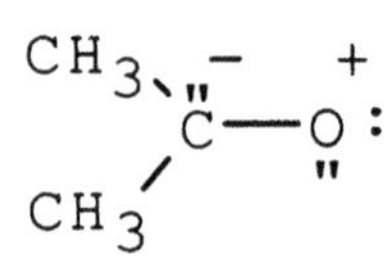

C

The molecule of acetone conforms to none of these individual electronic structures, but adopts a form, known as a **hybrid**, which reflects all of them to a degree. Each form contributes to an extent dependent upon its energy; the higher the energy of a form relative to the others, the less it contributes to the hybrid. Forms of too high an energy make a negligible contribution which we ignore.

? *Of the three forms so far written, which one would you expect to be of an energy much higher than the other two?*

* *

The answer is:

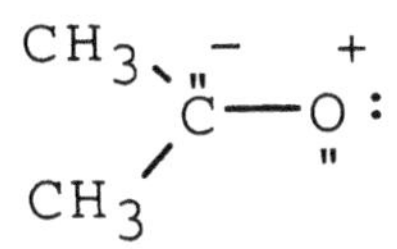

C

3

Whilst oxygen forms part of many stable structures in which it bears a negative charge (e.g. HO^-, $C_2H_5O^-$*), positively charged oxygen is usually only stable in certain situations (see part II, frame 2). To convert (in a 'thought experiment') the hypothetical form A into the hypothetical form C, a considerable input of energy would be required. To convert A into B, however, would require much less.*

? *Which element has the greater electronegativity, oxygen or carbon?*

* *

Oxygen has the greater electronegativity (Pauling Scale: C, 2.5; O, 3.5)

We can represent the 'thought conversion' of A into B, which requires the movement of a pair of electrons, using a 'curly arrow'. We write the curly arrow on the structure of one of the hypothetical forms (A), to represent the impending electron pair movement, and then write the outcome of this movement as though it were the product of a chemical reaction. It is just like writing a chemical equation, although it does not represent a real reaction or a real change. Hence the symbol ↔ rather than →.

4

A B

Note:

(i) The more electronegative oxygen atom will attract the π-electron pair more than carbon does, and that the shift is consistent with this attraction.

(ii) The curly arrow starts where the π-electron pair is in the π-bond, and the arrow head points clearly to the oxygen atom to which it is going.

? *Make a similar representation, using a curly arrow, for the 'thought conversion' of A to C.*

* *

You should have written:

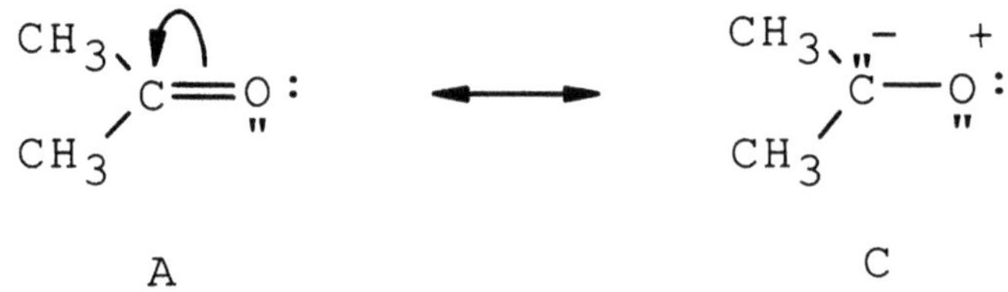

Note that this time the movement is against the direction favoured by the electronegativities of carbon and oxygen.

Since form C is of so high an energy relative to A and B, it makes a negligible contribution to the hybrid and we ignore it. We represent the hybrid, using forms A and B as 'adjectives', as follows:

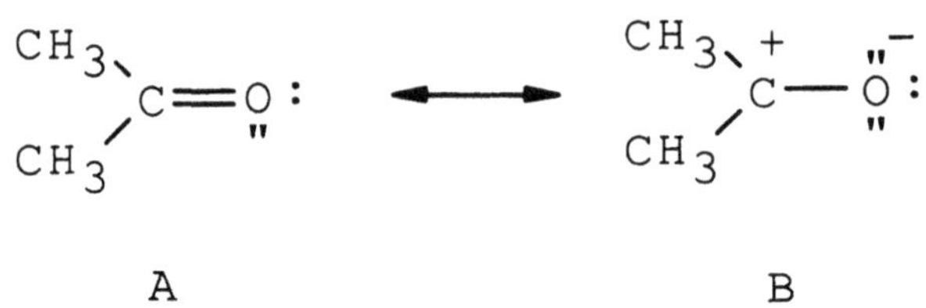

5 Note, again, the use of the double headed arrow, ↔, which should not be confused with the equilibrium symbol ⇌. The two forms are known as **canonical forms** (sometimes as **resonance forms**) and the molecule, a single species, is known as a **resonance hybrid**.

If we wish to write the hybrid as a single species, then we would do so as follows (in practice, rarely done for simple carbonyl compounds):

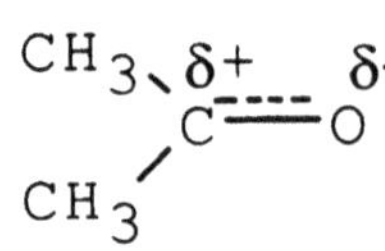

This shows that it is a sort of weighted 'average' of A and B, and that the carbonyl group is polar. ONLY THE HYBRID FORM EXISTS – CANONICAL FORMS HAVE NO EXISTENCE AT ALL.

The term 'resonance' is clearly rather unsatisfactory. It seems to imply that the molecule 'oscillates' between the various canonical forms, whereas this is not the idea at all. It is always a single species showing, in various degrees, the character of all of the canonical forms all of the time. An ass is a hybrid between a horse and a donkey. It is an ass all of the time, not a donkey on Mondays, Wednesdays and Fridays, and a horse on Tuesdays, Thursdays, Saturdays and Sundays!

? *Write valid canonical forms for formaldehyde (methanal), as well as a single representation of the hybrid.*

* *

You should have written:

canonical forms

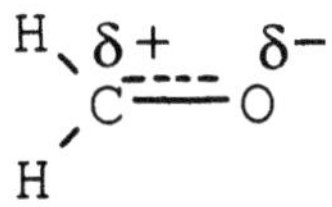

hybrid

If we had started by writing form **B** for acetone, then we could have predicted form A by using our curly arrow 'thought experiment'(see frame 4) the other way round:

6

B A

This emphasizes the involvement of lone pairs as well as π-electrons and illustrates an important rule for using curly arrows in this way. An electron pair may be moved from an atom to an adjacent bond or *vice versa.* We shall see later that it may also be moved from a bond to an adjacent bond. But, for any one shift, it always remains associated with one of the atoms in whose octet it 'started'.

? *Derive, with the help of curly arrows, resonance forms for the following ions.*

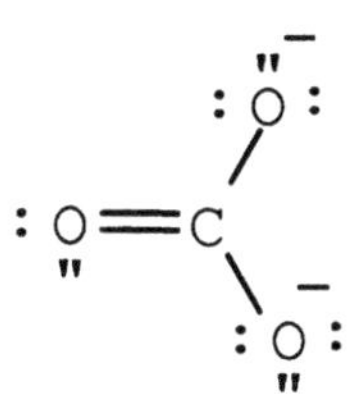

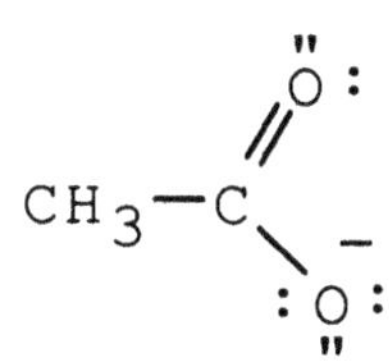

carbonate acetate

* *

You should have written:

carbonate

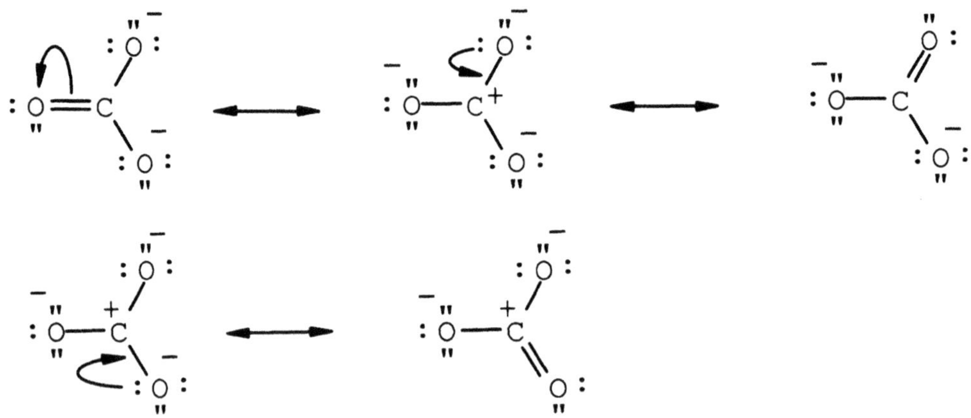

(second form repeated to show derivation of fourth form)

acetate

7

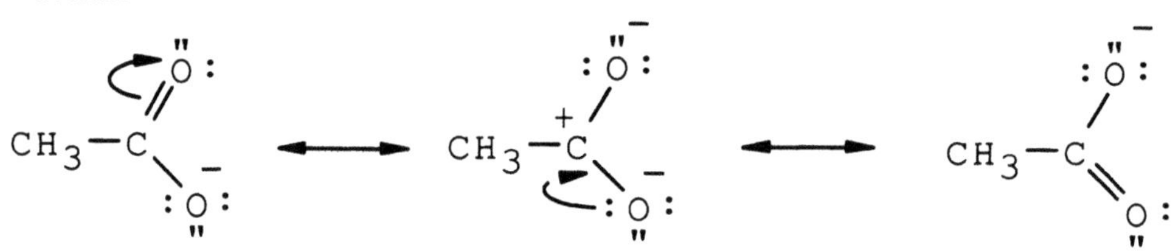

E F G

If your curly arrows are not placed exactly as they are here, with the tails and heads of the arrows in precisely the correct places, then re-write your forms before proceeding.

The three electron pairs which, in our 'thought experiments', we are moving about in the carbonate ion, and the two pairs in the acetate ion, are said to be **delocalized**. This concept is known as **delocalization** or, sometimes (unfortunately, see frame 5), as **resonance**.

Where we have several canonical forms, as with acetate, our curly arrows, if we link them, can take us between <u>any</u> of them. For example:

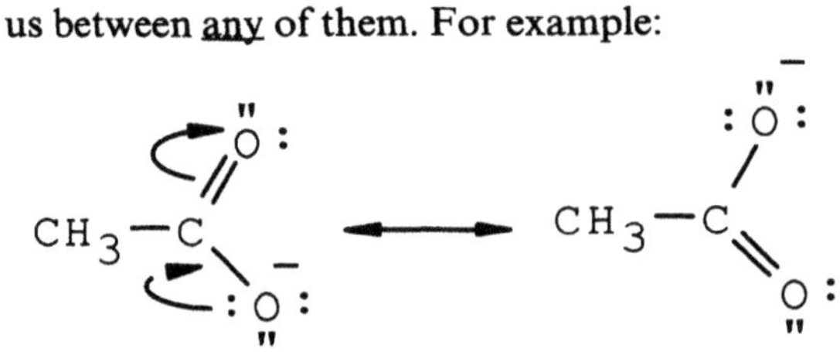

E G

We do not have to go through:

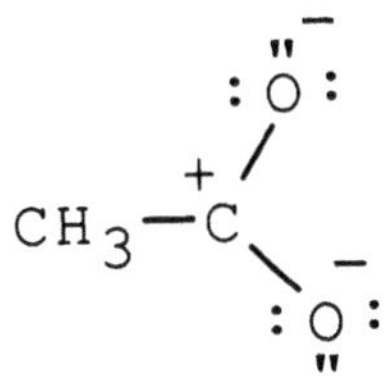

F

7

? *Give a single, linked curly arrow 'thought transformation' between:*

* *

You should have written:

8 Note that we cannot move an electron pair in from the $-O^-$ before we move one out from the double bond on to oxygen, since, if we did, we would break the octet rule and produce a form of impossibly high energy:

This form is inadmissible since the octet rule has been broken at carbon, and we have 5-valent carbon.

? *Would you expect form D below to be a valid resonance form for acetone?*

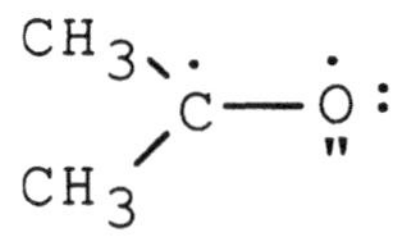

D

* *

No, because it has involved the unpairing of electrons, which 'costs' energy.

A rather special feature of the acetate canonical forms (see frame 7) is that two of them are clearly equivalent, and therefore are of identical energy. They are said to be **degenerate**. Their degeneracy must mean that they make an equivalent contribution to the hybrid. Since the 'conversion' of E into F involves the separation of opposite (i.e., attractive) charges, work must be done and form F will be of higher energy than form E. It will thus

9 make a smaller contribution to the hybrid. A good approximation to the hybrid would therefore be:

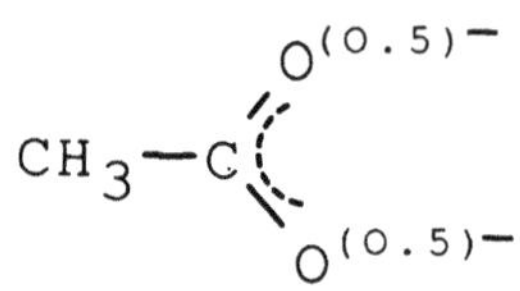

? *Which forms would be the principal contributors to the carbonate hybrid?*

* *

The correct answer is:

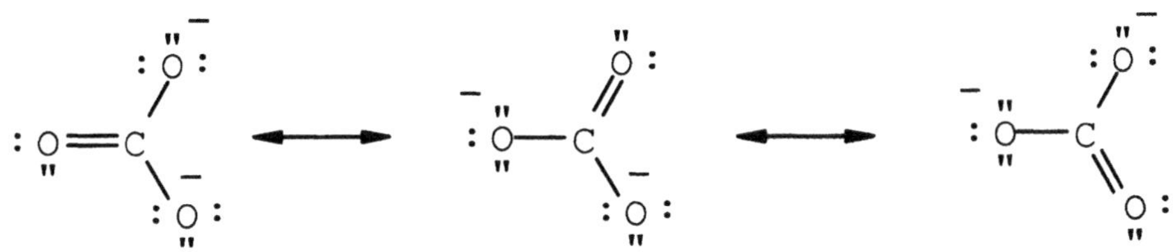

10

A hybrid form is always more stable than would be any of its individual canonical forms, if they could exist. Symmetrical resonance, involving equivalent forms making equal contributions, is the most effective sort of resonance and produces correspondingly greater stabilization of the hybrid.

Where there are several possible forms, the contribution of any particular form drops as the difference in energy between it and the lowest energy forms increases, so that very high energy forms make negligible contributions.

Thus, if there were, say, two possible forms, one (X) of very much higher energy than the other (Y), then X would make no significant contribution to the hybrid and the molecule would be no more stable than Y. If both of these forms were of identical energy, then each would make a 50% contribution to the hybrid, and the molecule (i.e. the hybrid), would be very much more stable than either of the separate forms.

It is important to be able to assess the relative contribution of each canonical form to the hybrid. The overall effectiveness of the resonance can then be appraised and any forms which are unlikely to make a significant contribution may be rejected.

? *Write possible resonance forms for the ethylene (ethene) molecule. How important would you expect resonance to be in this molecule?*

* *

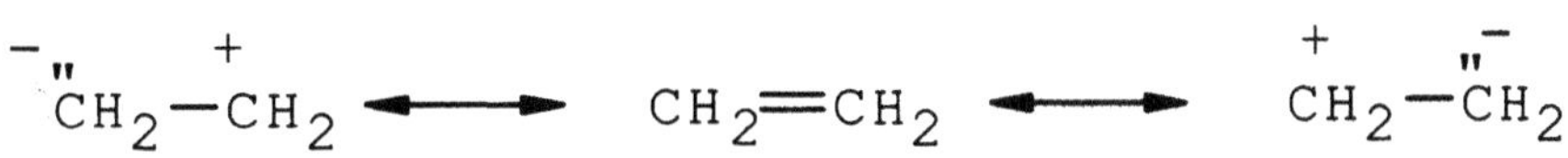

11 *Carbon (unlike oxygen) is reluctant to carry a negative charge if it can avoid it (or a positive charge), so that the dipolar forms are of far too high an energy, relative to CH_2=CH_2, to make any significant contribution. They will be much less likely to contribute here than does form B in acetone (see frame 12, below).*

Resonance is negligible here

? *In which species, acetone or acetate ion, would you expect delocalization to be most effective?*

* *

Acetate ion is the correct answer, since it has two equivalent canonical forms (see frame 7). Acetone (frame 5) has two fairly different forms; form B, which involves the separation of unlike charges and has a carbon with an incomplete octet, is likely to make only a small contribution.

Have you noticed that, in our 'thought experiments', movement of a <u>pair</u> of electrons results in the creation, or removal, of <u>single</u> charges on some of the atoms involved? Consider acetone again, for example:

A B

12 The reason for this, of course, is that the electron pair which we have 'moved' was only half owned by oxygen, and, when oxygen gains complete ownership, it gains a single negative charge. It acquires an extra half share in a pair. Similarly, carbon loses a half share in a pair of electrons and acquires a single, positive charge.

? *Using curly arrows, derive valid resonance forms for the nitrate ion and for nitromethane:*

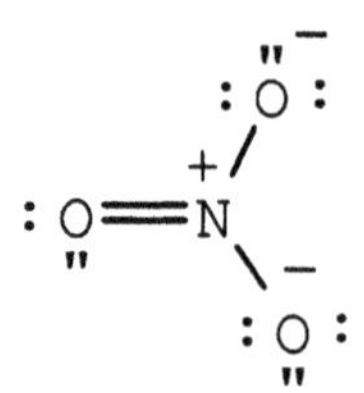

nitrate ion

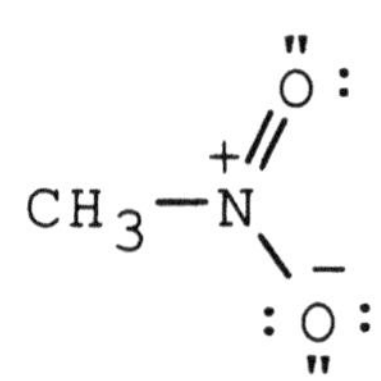

nitromethane

* *

The correct answers are:

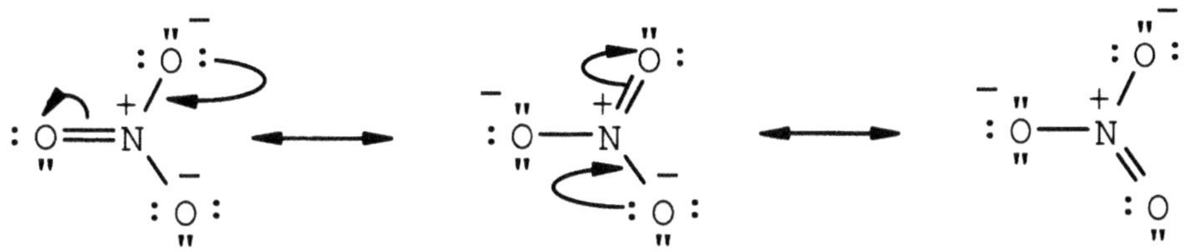

13 Again, we have examples where canonical forms are equivalent and delocalization is highly effective. Drawing the hybrid leads to the conclusion that, in practice, all three oxygen atoms of the nitrate ion are equivalent and should be experimentally indistinguishable.

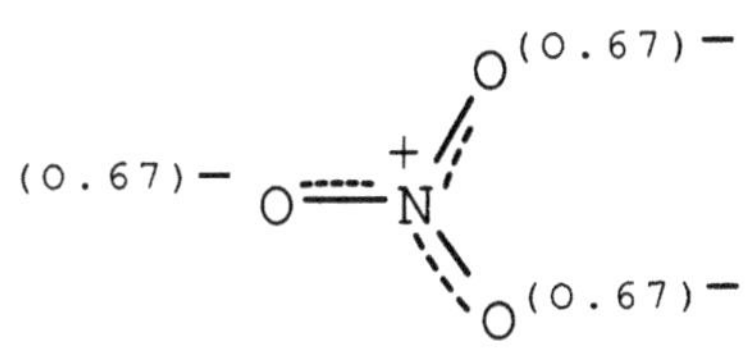

Experiment bears this out. The N to O bond lengths in this ion are all equal, shorter than a single N–O bond and longer than a double N=O bond.

? *(i) Draw a representation of the nitromethane hybrid and show that the two oxygen atoms are equivalent.*
(ii) Quote two other species from earlier frames which contain several atoms which, because of delocalization, would be experimentally indistinguishable.

* *

(i) What you should have drawn is:

14 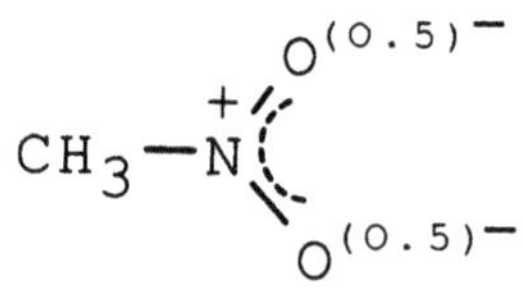

(ii) The acetate ion (frame 7) has two indistinguishable oxygen atoms, with equal C to O bond lengths, and the carbonate ion (frame 7) has three.

Review frame

It would be useful, at this stage, to review the essential rules to which resonance forms must conform in order to be significant contributors to a resonance hybrid (to be valid), and which have been covered in earlier frames.

I. All forms must have the same distribution of nuclei in space, i.e. the same geometrical arrangement of atoms.
II. No form may have an unacceptably high energy. In particular:
(a) no form may contravene the octet rule for those atoms which have atomic number of ten or less (this includes C, H, O, N and F);
15 (b) all forms must contain the same number of electron pairs (i.e. they must all contain the same number of electrons *and* these must be similarly paired in all forms).
III. All forms must carry the same total charge.

The primary advantages of using curly arrows for deriving resonance forms are that they help us to obey rule II(b) above, and to see, quickly, whether the movement is consistent with electronegativity differences. If we simply take a structure and strip it of π-electrons and lone pairs, and then feed these back into the structure in a different way, then there is a risk that some of the electrons may get 'lost'. In short, curly arrows help us with the electronic accountancy.

* *

This is the end of Part I of resonance and delocalization. Before attempting Part II (Programme 2) make sure that you are confident about the content of Part I (Programme 1). If you have any difficulty with these concepts – now or at any other stage – discuss them with your fellow students, or with a tutor.

Programme 2

Introduction to resonance and delocalization – Part II

Now that you are familiar with the basics of the concept of resonance (or, better, delocalization) we can extend the idea a little and apply it to more complex systems.

? *Derive, with the help of curly arrows, resonance forms for:*

1

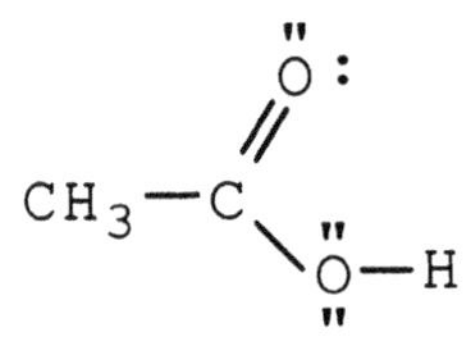

* *

You should have written the following forms:

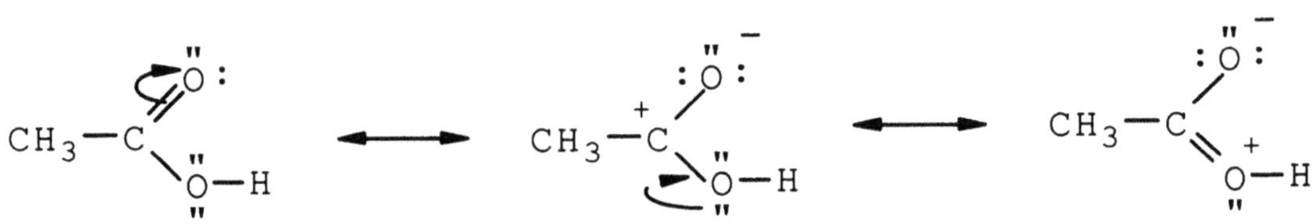

Note that, in spite of its electronegativity relative to carbon, the oxygen has, in one of these forms, acquired a positive charge, with electrons moving away from it. There are, of course, no forms possible in which that oxygen atom could bear a negative charge,
2 since that would break the octet rule. For example, form A below has ten electrons round the hydroxyl oxygen and is inadmissible.

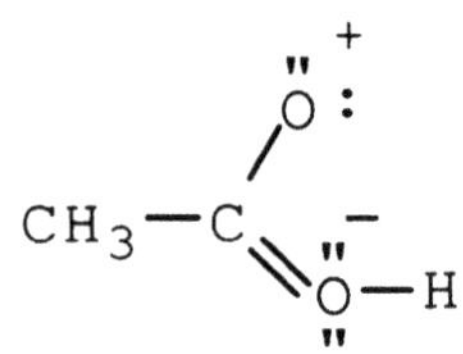

A

The only way in which this oxygen can delocalize a lone pair is to allow it to move towards carbon. It turns out to be energetically worth its while to 'do' this. Remember that when we rejected the $>\overline{\mathrm{C}}-\overset{+}{\mathrm{O}}$ form for acetone, there was an alternative, $>\overset{+}{\mathrm{C}}-\overline{\mathrm{O}}$, with a negative charge on oxygen (see Part I, frames 3 to 5).

? *Derive resonance forms for:*

(i) 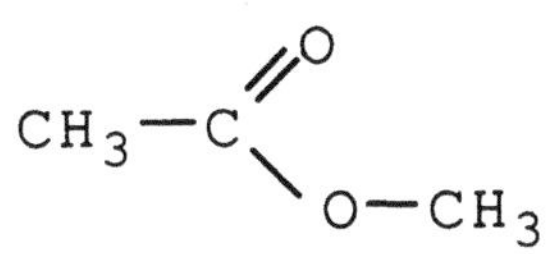 (ii) 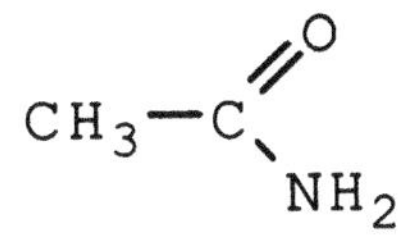

* *

You should have written the following forms:

(i) 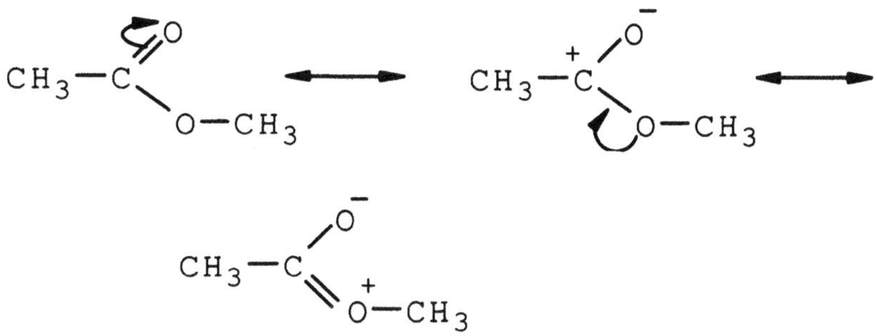

3

(ii) 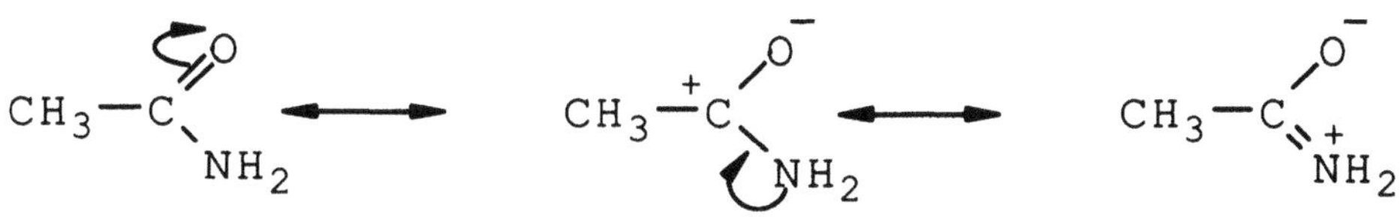

Note that we have no longer put lone pairs on the structures. They are still there of course, and we are still 'moving them about'. However, having reached this stage there is no longer any need to draw them in; they are taken as read.

? *Derive resonance forms for*

$CH_2{=}CH{-}O{-}CH_3$

* *

You should have written the following forms:

Remember that the 'curly arrow' represents the 'movement' of a lone pair (not shown) from oxygen to the adjacent (left hand) bond, and note that we have produced a negative charge on carbon. The dipolar form that we have drawn may make only a small contribution to the hybrid, but it is still important.

Another molecular site into which electron pairs can be delocalized is cationic carbon (which has only six electrons in its outer shell). Thus we can write canonical forms for the allyl cation as follows:

4

Note that once again we have two equivalent (degenerate) forms, and extremely effective delocalization. As well as delocalizing electrons here, we can also claim that the positive charge has been delocalized.

? *Write resonance forms for the species:*

(i) $CH_3{-}\overset{+}{C}H{-}O{-}CH_3$ (ii) $\overset{+}{C}H_2{-}CH{=}CH{-}O{-}CH_3$

* *

You should have written the following forms:

(i) 

(ii) 

Leading on from these examples, we find that, as well as moving electron pairs from atom to adjacent bond or vice versa (Part I, frame 6), we can also move them from bond to adjacent bond. Again, though, any pair, for any one shift, remains associated with one of the atoms with which it was previously associated. Thus, for propenal we can write:

5

Note that we can reject $\overline{C}H_2 - CH{=}\,CH{-}\overset{+}{O}$ for the same reasons that we rejected the $\rangle\overline{C}{-}\overset{+}{O}$ form for acetone (Part I, frames 3 to 5).

? *Derive resonance forms for benzene from the Kekulé structure:*

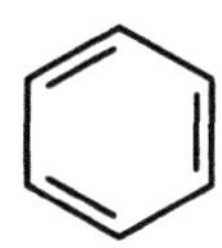

* *

The two most important forms are:

6

These forms probably have a very familiar ring to you; you will remember that the C to C bond lengths in benzene are all equal, and that benzene is a regular hexagon. They also fit in with our developing idea of delocalization and lead on to the question of the interaction between aromatic rings and their substituents.

? *Derive resonance forms for aniline (phenylamine), $C_6H_5NH_2$, remembering that there is a lone pair on the nitrogen atom.*

* *

You should have written the following forms:

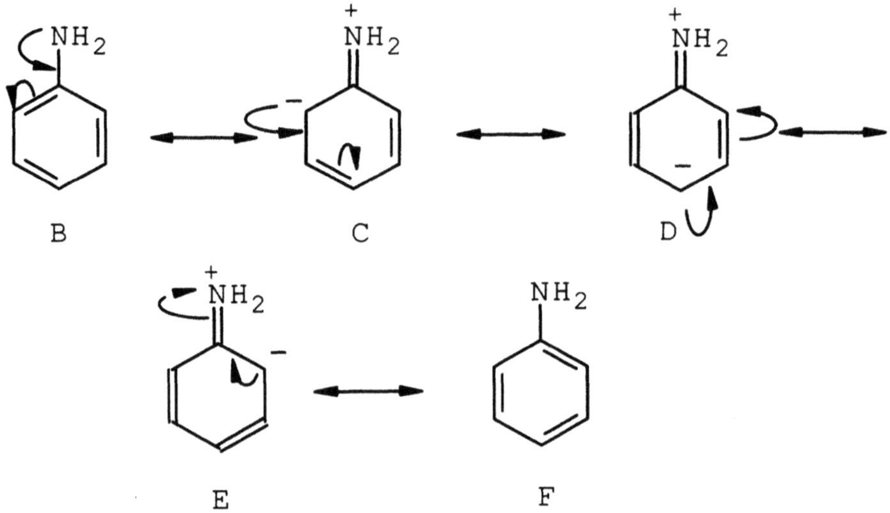

7

We have both Kekulé forms (B and F) as well as three others which tell us something about the interaction between the aromatic ring and the amino group substituent. Since we tend to take the delocalization within the ring for granted, we may not always bother to write form F.

Note very carefully that, in the shift from B to C for example, we cannot move the electron pair down from nitrogen without moving the π-bond pair on to C2, either first or at the same time. If we do not move the π-pair, there will be 'no room' for C1 to accommodate the nitrogen lone pair; we would be trying to break the octet rule at C1.

? *Derive resonance forms for:*

(i)

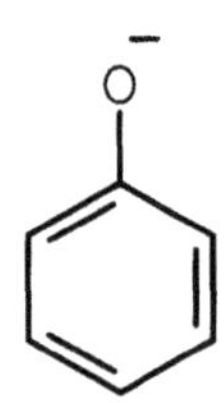

(ii)

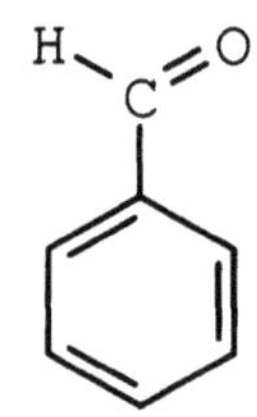

* *

You should have written the following forms:

(i)

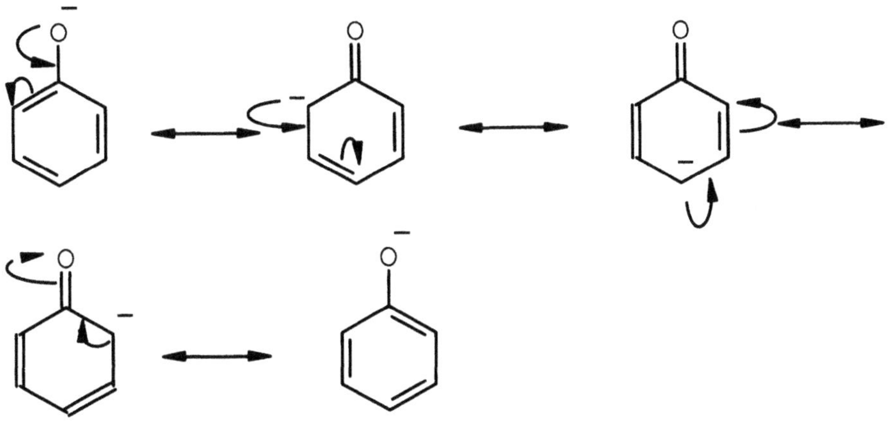

The case of this phenoxide ion is very like that of aniline, but charge (negative) is delocalized here, whereas in aniline unlike charges have been separated. Since the separation of unlike charges 'costs' energy, there is going to be a much greater difference in energy between the uncharged and the dipolar aniline forms than between any of the phenoxide ion forms. Delocalization is therefore likely to be more effective in the phenoxide ion than in aniline.

8

(ii)

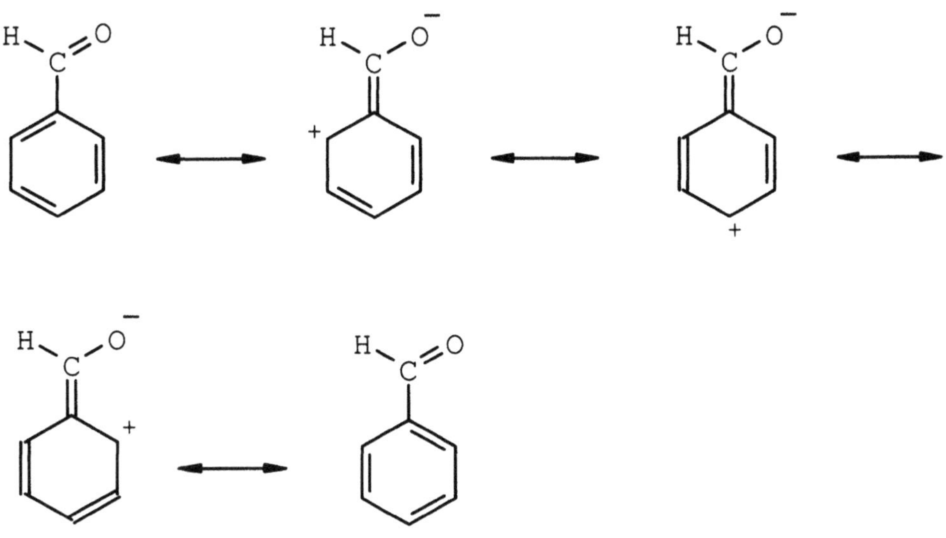

Here we note that the carbonyl group is able to withdraw electrons from the ring, whereas, by resonance, the amino group (in aniline) and the –O⁻ group (in the phenoxide ion), can only donate electrons to the ring.

Groups such as carbonyl, which withdraw electrons, are called **electron withdrawing** groups; groups which donate electrons, as the $-NH_2$ and $-O^-$ groups do in the examples of frames 7 and 8, are called **electron donating** groups. Sometimes this delocalization effect is referred to as the **mesomeric**, or **M effect**, and electron donors are said to be **+M** substituents, and electron withdrawing groups are said to be **–M** substituents.

The mesomeric effect should not be confused with the **inductive effect** which can operate in single or double bonds, and does not require lone or π-electron pairs. The inductive, or **I effect**, is usually related to the differences in electronegativity between the atoms at either end of a bond, and may operate in the opposite direction to any M effect. Thus the amino group has a –I effect – nitrogen is more electronegative than carbon – and a +M effect.

? *Classify the following groups according to their likely M and I effects when they are attached to a benzene ring, i.e. say whether they are likely to show a + or a –M effect, or no M effect at all, and similarly, what their I effect is likely to be.*

$-Cl$; $-OH$; $-O-CH_3$; $-N(CH_3)_2$; $-CO-CH_3$ $-C{\equiv}N$; $-\overset{+}{N}(CH_3)_3$

* *

The correct classifications are:

	$-Cl$;	$-OH$;	$-O-CH_3$;	$-N(CH_3)_2$;	$-CO-CH_3$;	$-C{\equiv}N$;	$-\overset{+}{N}(CH_3)_3$
M	+	+	+	+	–	–	nil
I	–	–	–	–	–	–	–

On all of the monosubstituted benzene derivatives which we have looked at, the delocalization has put charges on positions 2, 4 and 6 of the ring (position 1 is defined as
9 the substituent bearing carbon atom), but not on positions 3 and 5. This **alternacy** of aromatic systems like benzene is very important.

? *Derive resonance forms for the two molecules below:*

(i)

(ii)

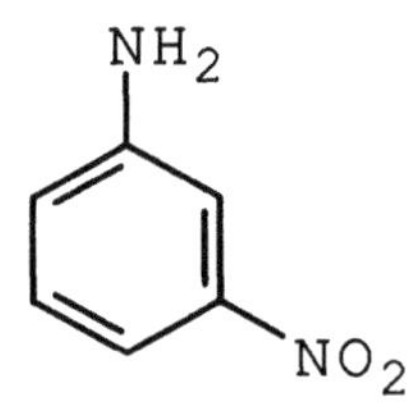

* *

You should have written the following forms:

(i) 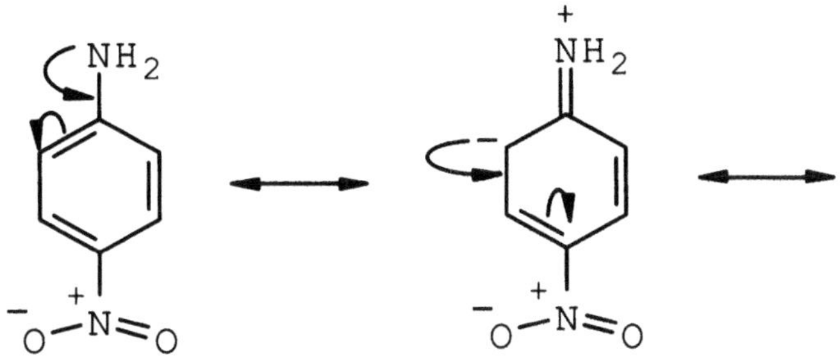

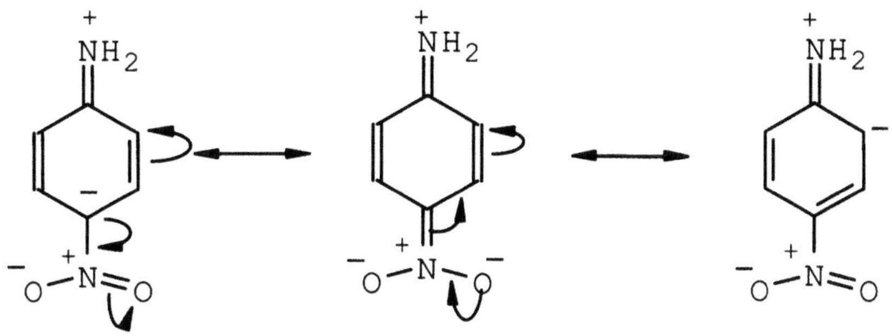

10

(ii) 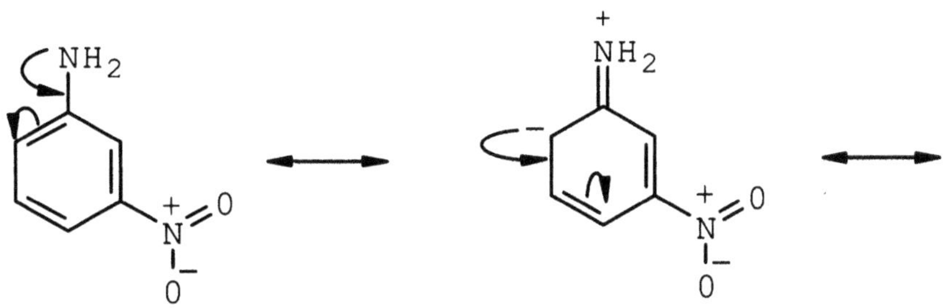

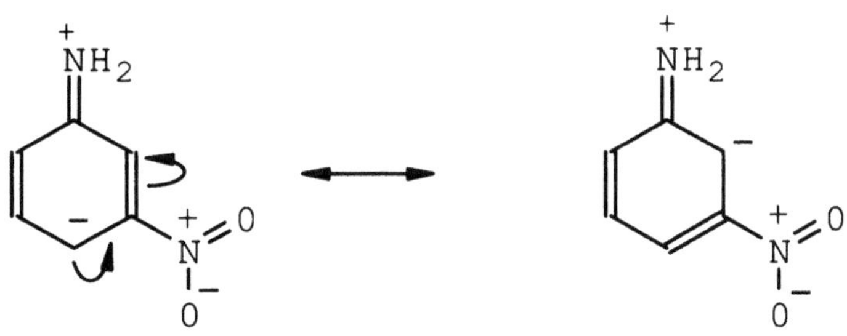

Note that there is no form for (ii) in which the amino nitrogen lone pair can be delocalized into the nitro group, whereas for (i) there is.

This is another reflection of the alternant properties of the benzene ring and has important chemical consequences; compound (i) is a weaker base than (ii), for instance, but we shall take this up in another programme (Programme 3).

? *Write resonance forms for:*

(i) 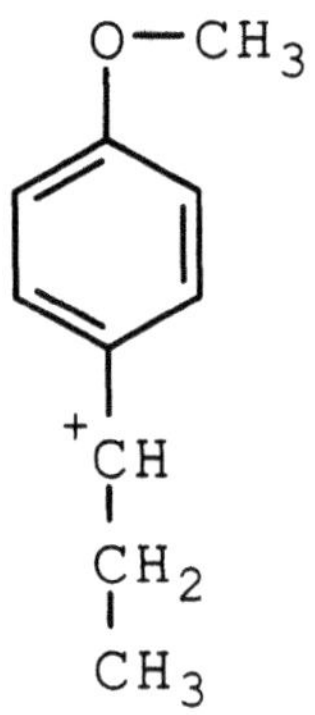 and (ii) 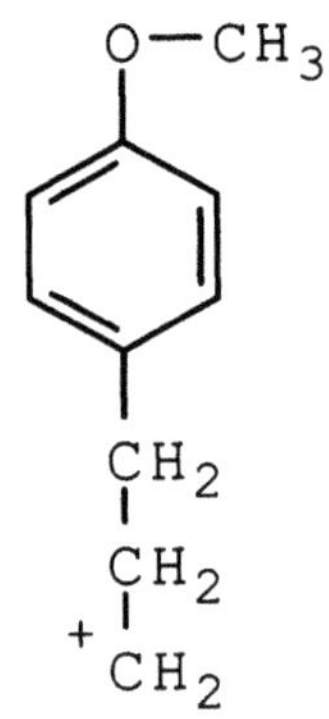

* *

You should have written the following forms:

(i) 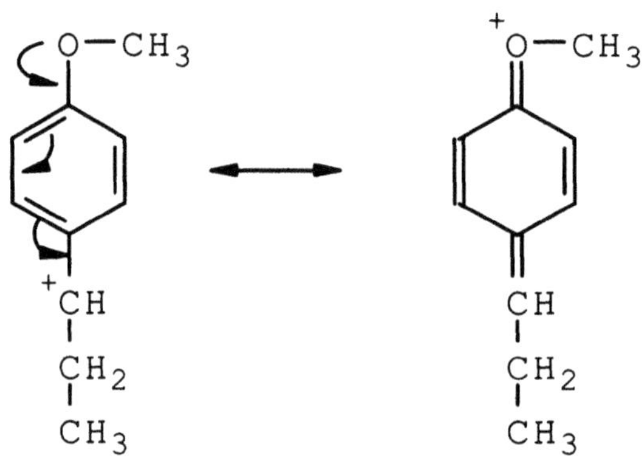

(and several others)

11

(ii) 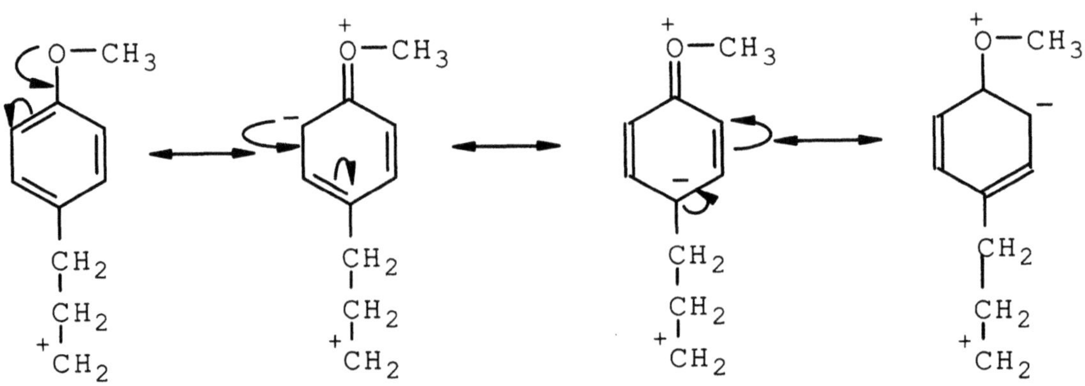

Here we can predict that the cation (i) is stabilized by resonance to a greater extent than is the cation (ii), since the positive charge in (i) is delocalized right out on to oxygen, whereas no delocalization is possible for the positive charge in (ii).

? *Which of the two cations below would be the more stable?*

(i) $CH_3{-}\overset{+}{C}H{-}CH{=}CH_2$ or (ii) $\overset{+}{C}H_2{-}CH_2{-}CH{=}CH_2$

* *

You should have written resonance forms as follows:

(i) $CH_3{-}\overset{+}{C}H{-}CH{=}CH_2 \longleftrightarrow CH_3{-}CH{=}CH{-}\overset{+}{C}H_2$

(ii) no analogous resonance forms are possible

From this one can conclude that (i) is more stable than (ii).

Frequently, application of the concept of resonance leads to conclusions of chemical
12 importance. Although this is the topic of another programme, we will consider one example here.

? *There are two different sites upon which buta-1,3-diene may be protonated:*

$$CH_2{=}CH{-}CH{=}CH_2 + H^+ \longrightarrow CH_3{-}\overset{+}{C}H{-}CH{=}CH_2 \text{ (C1 protonated)}$$

or

$$\overset{+}{C}H_2{-}CH_2{-}CH{=}CH_2 \text{ (C2 protonated)}$$

Which of the two do you think would be preferred?

* *

Protonation at C1 would be preferred, since the cation so produced (with charge at C2) can be stabilized by resonance, whereas protonation at C2 leads to a cation (charge at C1) which cannot be resonance stabilized:

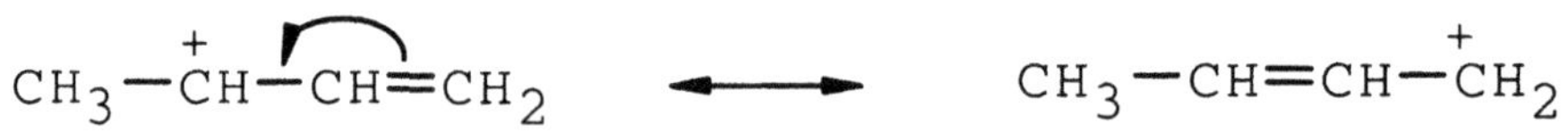

13

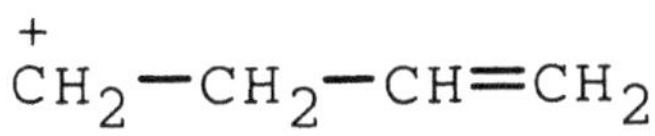

no analogous resonance forms are possible

? *Which of the two compounds below would be the more stable?*

* *

Compound (i) would be the more stable, since it can be stabilized by resonance:

14

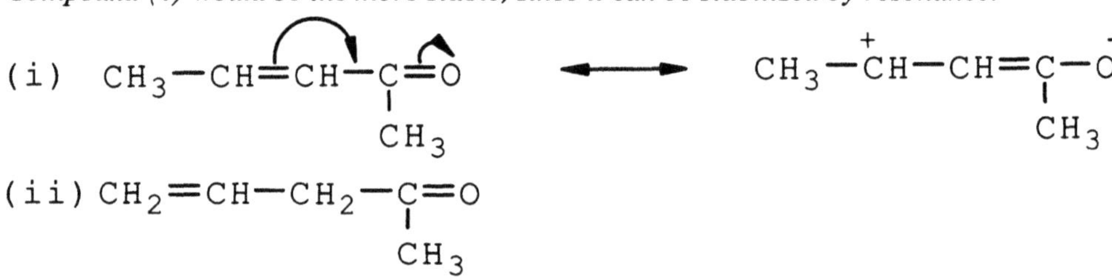

no resonance forms possible

(beyond those for the isolated C=O)

Systems with alternating double and single bonds are described as **conjugated** systems, and are stabilized by delocalization. Two adjacent single bonds (i.e. an sp^3 atom) between two double bonds prevent the double bonds from being conjugated, as in compound (ii) above. The absence of any single bonds between two double bonds will also prevent those double bonds from being conjugated. Double bonds in this situation (i.e. which are adjacent) are said to be **cumulated**, as in ketene or allene:

ketene: $CH_2{=}C{=}O$ allene: $CH_2{=}C{=}CH_2$

? *Satisfy yourself, by trying to draw appropriate resonance forms, that there can be no resonance interaction between the double bonds of ketene or those of allene.*

* *

ketene 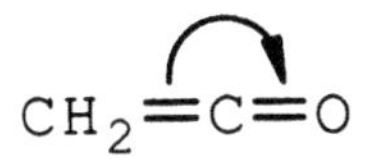

forbidden – oxygen cannot expand its octet

allene 

forbidden – carbon cannot expand its octet

This absence of interaction between two cumulated double bonds is examined further in Programme 3, frames 25 and 26.

In establishing the principles of resonance, we have considered only species containing an even number of electrons, so that the electrons can always be paired. Species containing unpaired electrons – usually containing an odd number of electrons – are known as **free radicals** and may also be stabilized by resonance. In writing resonance forms for these species, though, we move electrons around singly. The rules for resonance forms (Part I, frame 14) do not change, and in particular, rule II(b) implies that resonance forms for
15 free radicals must contain the same number of unpaired electrons (as well, of course, as the same number of electron pairs!). To represent the shift of a single electron we use a curly arrow with only half a head and the unpaired electron is shown as a single dot:

? *Write resonance forms for the following free radicals:*

(i) $CH_2{=}CH{-}CH{=}CH{-}\dot{C}H_2$

(ii) 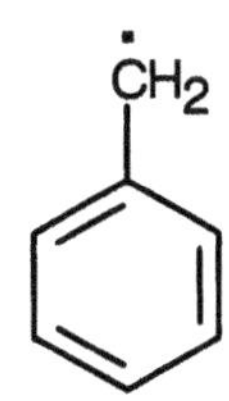

* *

You should have written the following forms:

(i)

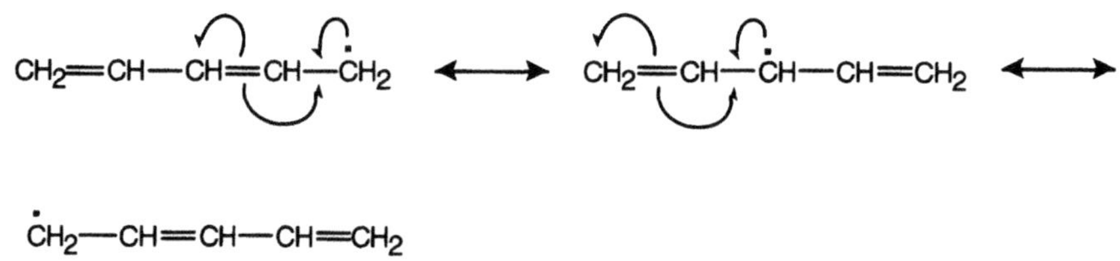

16

(ii)

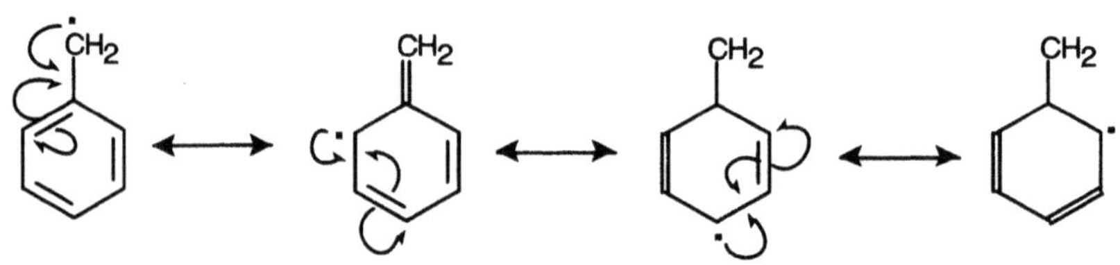

This is the end of this programme on resonance and delocalization. When you are quite confident that you have mastered this material then you can proceed to the more applied programme on the role of electronic effects in determining the strengths of acids and bases, Programme 3. To help ensure that you have mastered resonance, some examples and solutions follow this programme. It is recommended that you do these examples, and check your solutions, before proceeding to Programme 3.

* *

Further examples and solutions

? *Write resonance forms for the following molecules and ions:*

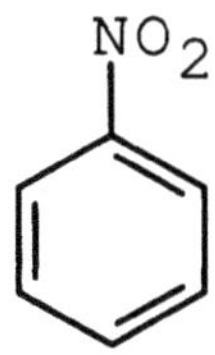

1.

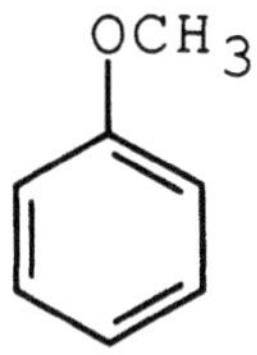

2.

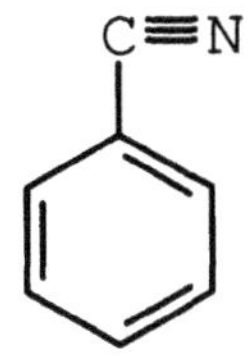

3.

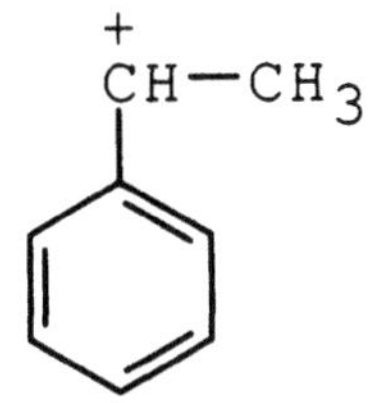

4.

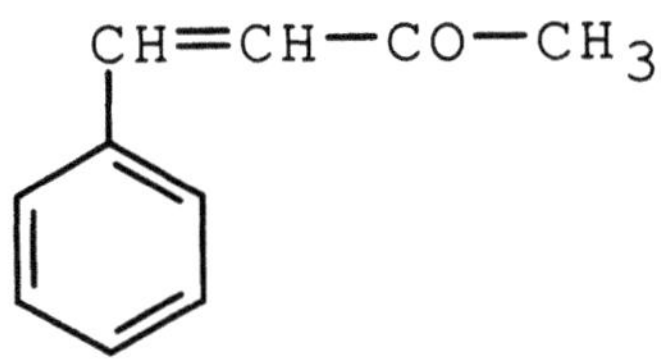

5.

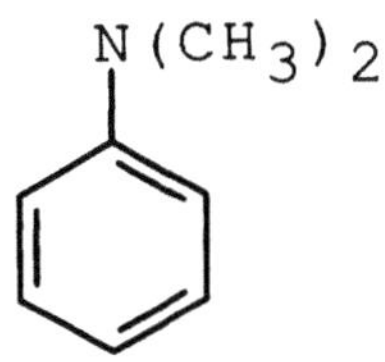

6.

$CH_3{-}CH{=}CH{-}CO_2CH_3$

7.

$CH_3{-}CO{-}CH{=}CH{-}OCH_3$

8.

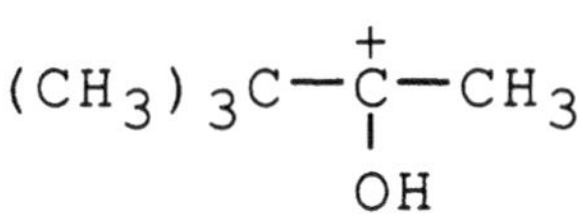

9.

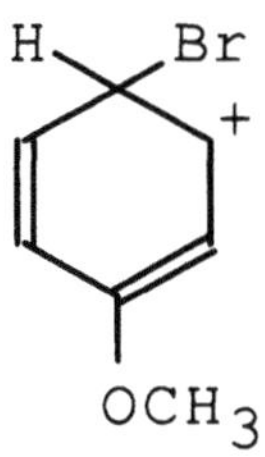

10.

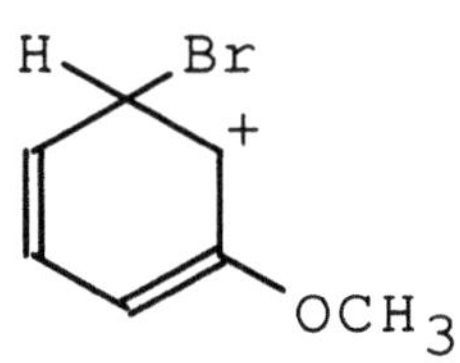

11.

* *

Solutions:

1. 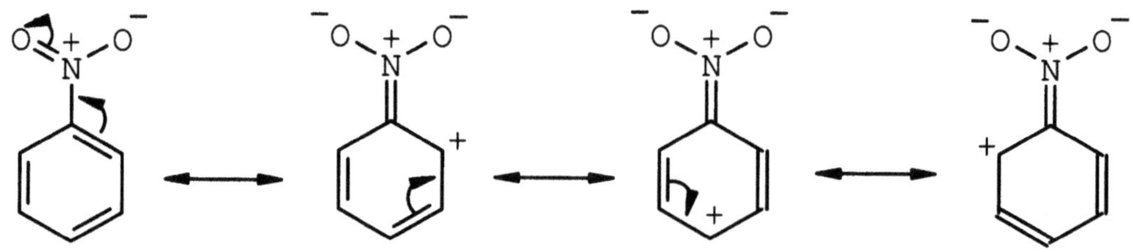

2. 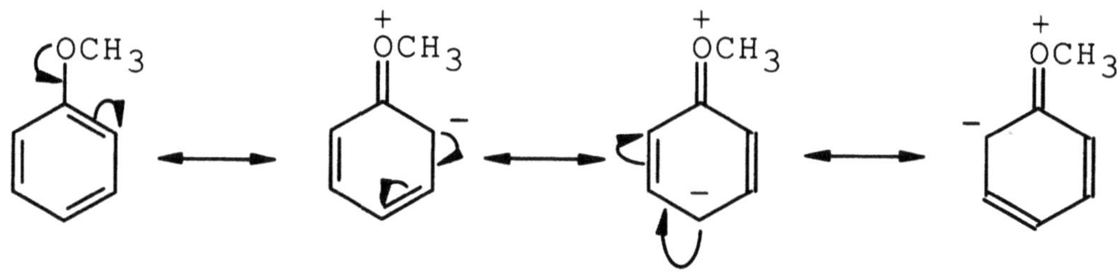

3. 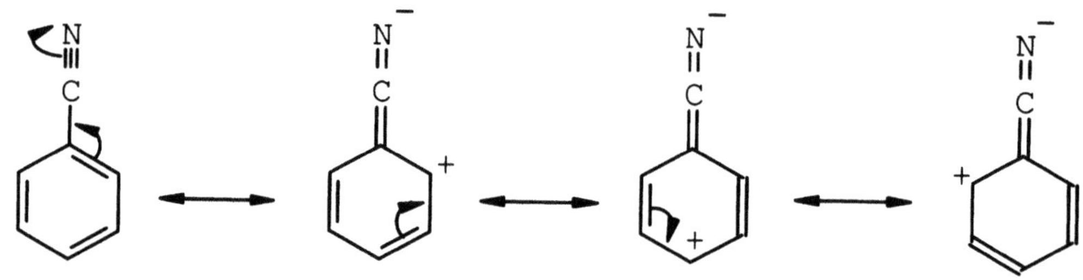

4. 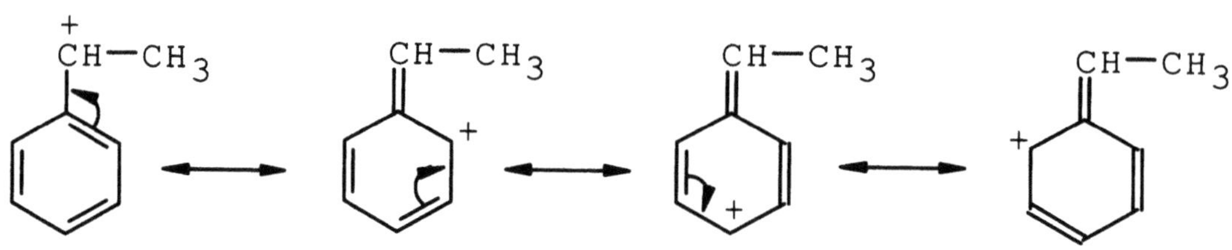

5.

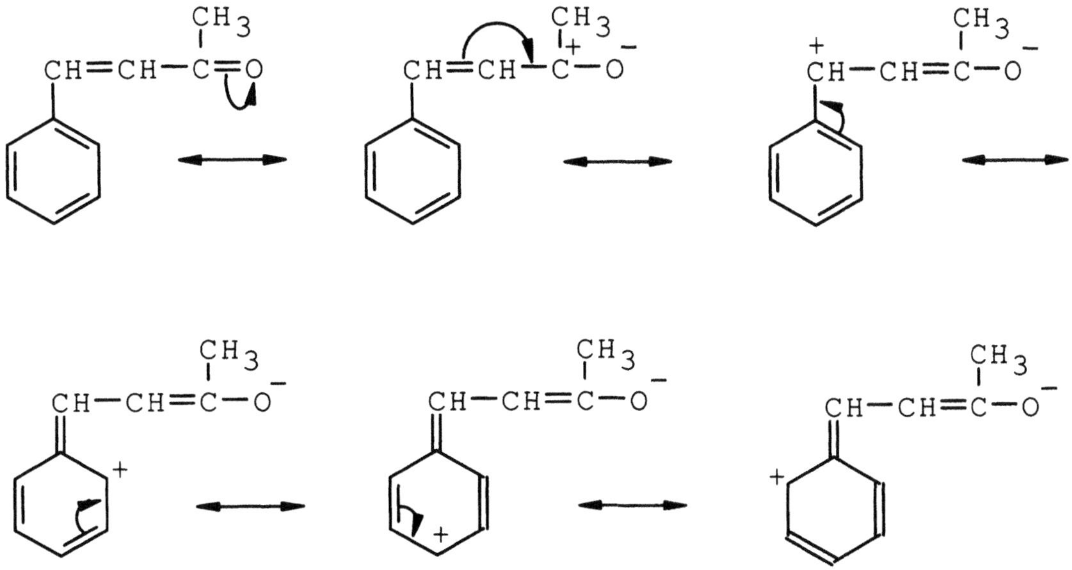

6.

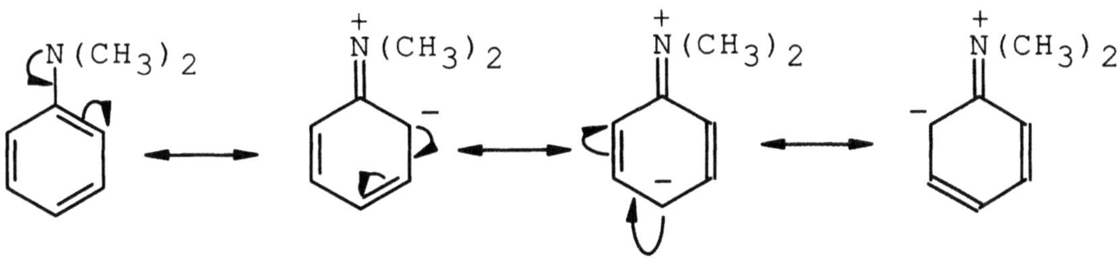

7.

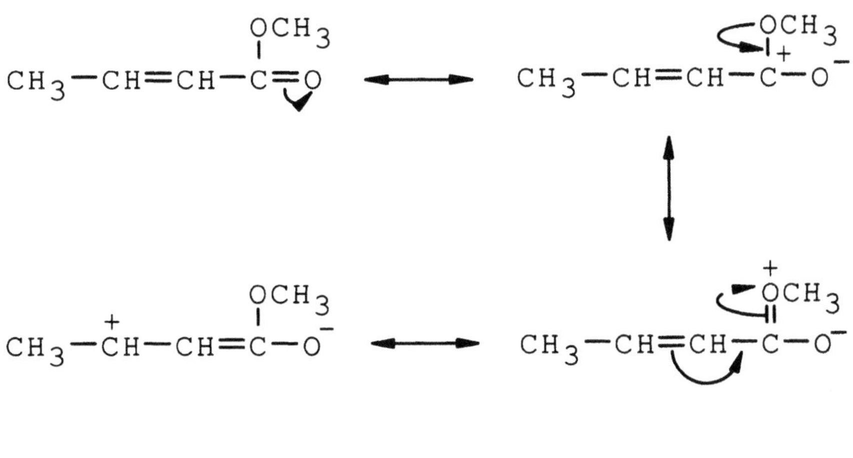

8.

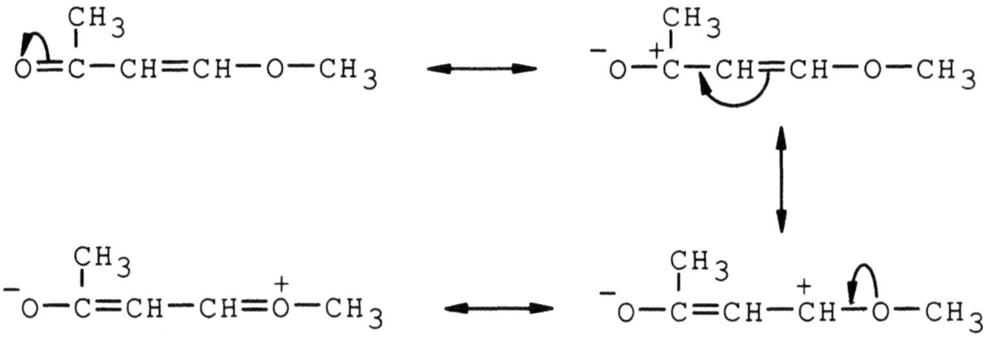

9.

10.

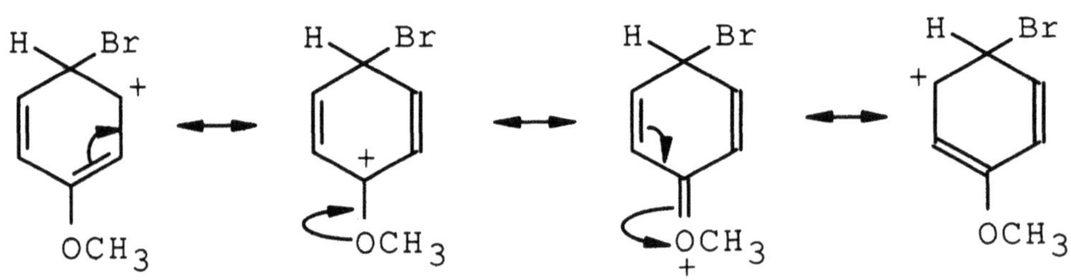

11.

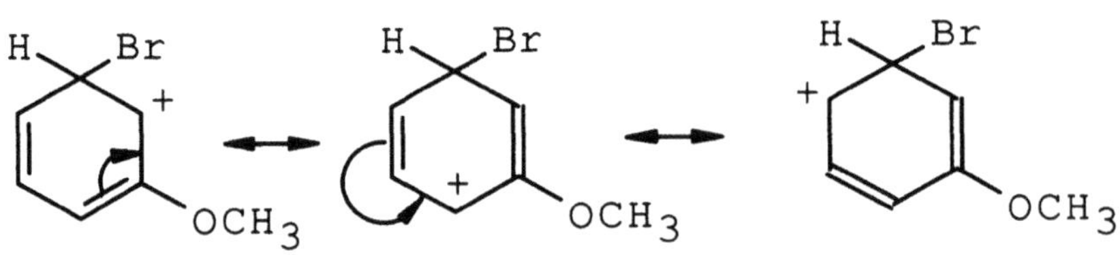

Programme 3

Strengths of acids and bases – electronic effects

In this programme we are going to examine the use of electronic effects – resonance and inductive – in explaining or predicting differences in strengths between various organic acids and bases. The programme begins by considering some of the basic concepts which enable us to express these strengths quantitatively, and some of the essential thermodynamic factors which determine their magnitudes.

When a simple Brönsted acid, HA, dissolves in water, an equilibrium is set up as follows:

$$HA + H_2O \overset{K_a}{\rightleftharpoons} H_3O^+ + A^-$$

1

The acid HA protonates a water molecule and the equilibrium constant for the reaction is K_a. All of the species will be solvated, of course, but for the moment we shall ignore that.

The magnitude of K_a is a measure of how far the equilibrium lies over to the right, i.e. how effective HA is at protonating water. We can therefore use K_a as a measure of the strength of an acid and if we compare K_a values for a variety of acids we have a quantitative comparison of how good they are at protonating a common base, water.

Frequently, for convenience, we express the strength of an acid as its pK_a value, $-\log_{10}K_a$. This means, of course, that **strong** acids have **low** pK_a values (just as strongly acidic solutions have low pH values), and it is important to remember this.

? *What is the fundamental thermodynamic quantity which determines the size of an equilibrium constant, and how is it related to the equilibrium constant?*

* *

2

The difference in standard free energy between reactants and products, $\Delta G°$, and this is related to the equilibrium constant by the expression:

$$\Delta G° = -RT\ln K_a$$

Where R is the gas constant and T is the absolute temperature.

So, if we are to make predictions about the strength of an acid, then we must make predictions about $\Delta G°$ for its reaction with water.

? *There are two other fundamental thermodynamic quantities which relate to $\Delta G°$ and which are sometimes useful for calculating $\Delta G°$. What are they, and how are all three quantities related?*

* *

They are the standard enthalpy and standard entropy differences between reactants and products, ΔH° and ΔS° respectively, and they are related by the expression (T = absolute temperature):

$$\Delta G^\circ = \Delta H^\circ - T\Delta S^\circ$$

The value of ΔG° for any acid/base reaction depends upon ΔH° and ΔS°, both of which will depend upon a variety of factors including solvent, electronic (resonance and inductive) and steric effects. Correlation of these effects with ΔH° and ΔS° is neither necessary nor possible at this stage, but it turns out that the use of electronic effects alone leads to good predictions of the strengths of acids and bases. This is clearly not 'theoretically rigorous', **but it works** (the 'acid' test of any technique!), and we therefore use it.

Consider two acids, HA and HA′. Each will react with water:

3

$$HA + H_2O \overset{K_a}{\rightleftharpoons} H_3O^+ + \bar{A}$$

$$HA' + H_2O \overset{K'_a}{\rightleftharpoons} H_3O^+ + \bar{A}'$$

Since we are concerned only with the difference in ΔG° between the two reactions – we might call this $\Delta\Delta G^\circ$ – then we can forget about H_2O and H_3O^+ since they are common to both reactions. Any factor then which makes $\bar{A}'$ more stable relative to HA′ than A^- is relative to HA, would suggest that ΔG° for the lower reaction is larger (more negative) than ΔG° for the upper reaction, and that K_a' is larger than K_a. In other words, HA′ is a stronger acid than is HA.

So, if $\bar{A}'$ is stabilized by resonance, whereas neither HA′, HA, nor A^- is, then we could conclude that HA′ should be the stronger acid.

? *Account for the fact that ethanol is a weaker acid than acetic acid.*

* *

The acetate ion is stabilized very effectively by resonance:

Neither ethanol (C_2H_5OH) nor the ethoxide ion (C_2H_5O)$^-$ are stabilized by resonance – we cannot draw any resonance forms for them – and although acetic acid itself is stabilized by resonance, this stabilization is not very effective compared with that in the acetate ion. In the latter, both negative charge and electrons are delocalized, and the resonance is highly symmetrical, whereas in the former the charged form has unlike charges separated and is of relatively high energy.

4

So, we can see why acetic acid may be expected to be a stronger acid than ethanol.

Similar considerations may be applied to the strengths of bases. In the characteristic base equilibrium in water, we are concerned with differences in stability between B and BH^+.

$$B + H_2O \rightleftharpoons BH^+ + OH^-$$

Electronic effects may enable us to predict that, say, B is a weaker base than B′ because the stability of B relative to BH^+ is greater than the stability of B′ relative to $B'H^+$.

? *Which would you predict to be the weaker base, cyclohexylamine, $C_6H_{11}NH_2$, or aniline (phenylamine), $C_6H_5NH_2$?*

* *

On the basis of resonance effects, aniline would be expected to be the weaker base since the lone pair on the aniline nitrogen atom can be delocalized into the ring, whereas when the amino group has been protonated, that pair of electrons is 'localized' in the new N–H bond. The stabilization associated with the delocalization of that lone pair is lost on protonation. No delocalization is possible in the cyclohexylamine molecule which therefore has no resonance stabilization to lose on protonation.

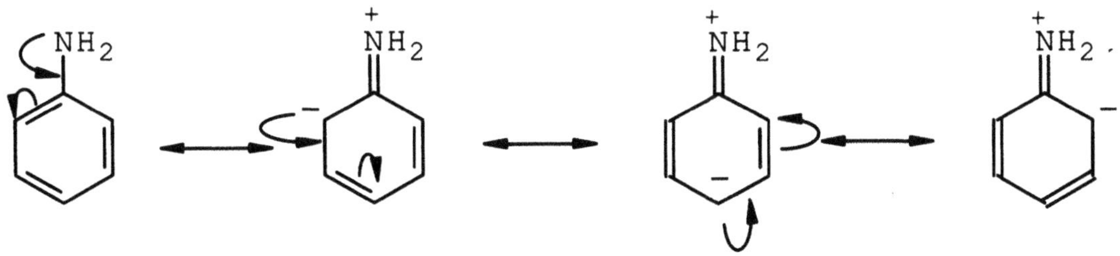

Compare:

5

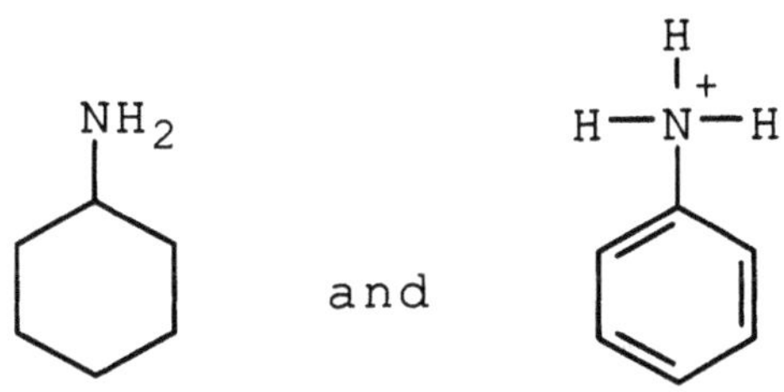

This prediction is born out by experiment. Aniline has a pK_a value of 4.63 and cyclohexylamine has one of 10.66.
Note here the rather surprising use of pK_a for bases rather than pK_b.

$$B + H_2O \overset{K_b}{\rightleftharpoons} BH^+ + OH^-$$

pK_a values are commonly used for both acids and bases, and this can be confusing if you forget it. What one is really quoting here is the strength of BH^+ as an acid. A strongly acidic BH^+ implies a weakly basic B, so that a low pK_a value for a base, B, means that BH^+ is a strong acid, and B is a weak base. **The higher the pK_a value of a base, the stronger it is, whereas the higher the pK_a value of an acid, the weaker it is.**

? *Which would you predict would be the stronger base, ethylamine or acetamide (ethanamide)?*

* *

Ethylamine should be the stronger base. No delocalization of the nitrogen lone pair is possible in ethylamine, but in acetamide the lone pair is extensively delocalized into the carbonyl group, pulled by the strongly electron withdrawing carbonyl group.

6

We do not have to confine ourselves to pairs of compounds; the groups which we consider can be much larger. We might, for example, consider the relative basicities of ethylamine, aniline and acetamide.

? *Arrange these three compounds in decreasing order of their predicted basicities at nitrogen.*

* *

The predicted order is:

ethylamine > aniline > acetamide

7

All three of these compounds have already been considered, and resonance forms drawn where appropriate (frames 5 and 6). It is fairly easy to predict that ethylamine, with no resonance forms, should be the strongest base. It is rather more difficult to decide about aniline and acetamide, since both are resonance stabilized, and both will lose the ability to delocalize their nitrogen lone pair when protonated. However, forms in which oxygen bears the negative charge, as in acetamide, are likely to be much more stable than forms in which carbon bears the negative charge, as in aniline. This difference is likely to be quite large, and is reflected in a large difference in basicity between acetamide ($pK_a = 0.63$) and aniline ($pK_a = 4.63$). Ethylamine is very different again ($pK_a = 10.8$).

Remember that a difference of one between two *pK* values represents a factor of ten between the equilibrium constants. A difference of two corresponds to a factor of one hundred, a difference of three corresponds to a factor of one thousand, etc. The fact that the pK_a values are so different here, and that the electronic effects are so very different too, is fortunate. With structures which are so different any assumptions that entropy and solvation are similar for them all, and may be ignored, are likely to be rather shaky, and we can probably only handle big differences. Finer tuning would be impossible with this qualitative approach.

The effect of one substituent upon another in an aromatic molecule requires the comparison of molecules which are very similar in structure, and which therefore make our discounting solvent and steric effects much more reasonable. Thus, we could make predictions about the acidity of nitrated phenols with rather more confidence.

? *Which would you expect to be the stronger acid, phenol or* p-*nitrophenol?*

* *

p-nitrophenol is expected to be the stronger acid, since the p-nitrophenoxide ion is more effectively stabilized by resonance than is the unsubstituted phenoxide ion.

Both the phenoxide ion and the *p*-nitrophenoxide ion will be resonance stabilized, with delocalization of a lone pair of electrons and the negative charge from the 'oxide' oxygen.

Phenoxide ion:

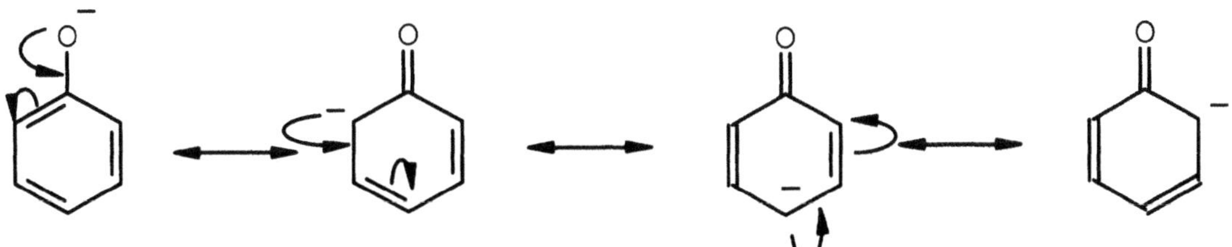

p-nitrophenoxide ion:

8

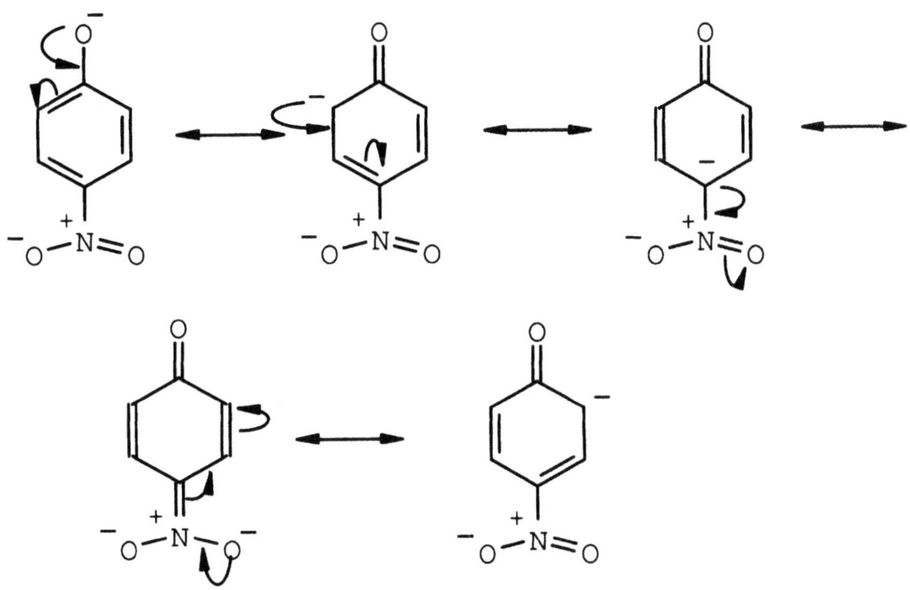

In the latter case, with the lone pair delocalized right down on to a nitro group oxygen, not only can we write more resonance forms, but on one of them the negative charge resides on an electronegative oxygen atom. In the phenoxide ion forms, apart from the 'oxide' oxygen itself, the negative charge can only reside on a carbon atom. We can thus expect stabilization to be greater for the *p*-nitrophenoxide ion.

The relevant pK_a values are:

phenol	9.89
p-nitrophenol	7.15

The stabilization of the unsubstituted phenoxide ion would also enable us to predict that phenol should be a stronger acid than ethanol (or any simple alcohol) (frame 4).

Although the structures of phenol and ethanol are not all that similar, the difference is between a case where resonance can stabilize an anion (phenoxide) and a case where it cannot help at all (ethoxide). The experimental difference in the pK_a values is also large. Relevant pK_a values, with that for acetic acid, are:

8

ethanol	16
phenol	9.89
acetic acid	4.75

? *Note the difference in* pK_a *values between phenol and ethanol (a factor of a million!), and note the position of acetic acid in this league. Why would you expect acetic acid to be stronger than phenol? (Hint – draw out the resonance forms for acetate and phenoxide ions and compare them. If you are uncertain about them, check them against frames 4 and 8).*

* *

9

The acetate ion has two exactly equivalent (degenerate) resonance forms, both of which carry the negative charge on oxygen. With a much less symmetrical form of resonance, and charge delocalized from oxygen only on to carbon, the phenoxide ion is less effectively stabilized than is the acetate ion, and acetic acid would therefore be expected to be the stronger acid.

A further pK_a value of interest at this point is that for *m*-nitrophenol, and this is 8.28.

? *Account for the observation that* m-*nitrophenol is a weaker acid than is* p-*nitrophenol.*

* *

The alternant nature of benzene (Programme 2 – frames 9 and 10) makes it impossible to delocalize the negative charge from the 'oxide' oxygen of the m-*nitrophenoxide ion on to a nitro oxygen, hence this anion is less effectively stabilized than is the* p-*nitrophenoxide ion. This would lead to the expectation that* m-*nitrophenol is a weaker acid than the* p-*nitrophenol.*

10

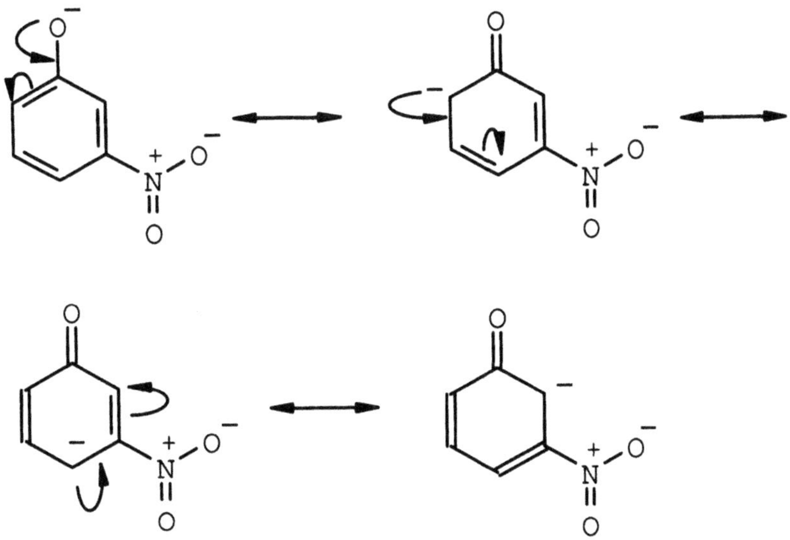

Try to draw a resonance form for this anion with the negative charge on a nitro group oxygen. If you succeed, test your proposed form very carefully against the resonance rules formulated in Part I of the resonance and delocalization programme (Programme 1, frame 15).

? *With a* pK_a *value of 8.28,* m-*nitrophenol is a stronger acid than phenol itself (9.89). Since the negative charge in the* m-*nitrophenoxide ion cannot be delocalized on to a nitro group oxygen, why should this be so?*

* *

The nitrogroup stabilizes the m-*nitrophenoxide ion by* **inductive** *electron withdrawal, so that* m-*nitrophenol is a stronger acid than is phenol.*

Although it is the case that the nitro group cannot accept the lone pair and negative charge from the 'oxide' oxygen of the *m*-nitrophenoxide ion, the nitro group is still withdrawing electrons from the ring. This it is doing by both a resonance effect and an inductive effect. The nitro group is a –M and a –I group (Programme 2, frames 8 and 9). It is, by the resonance effect, placing a positive charge on the positions *o*– and *p*– to it, and to 'see' this we need to draw out resonance forms which reflect the interaction between the nitro group and the ring:

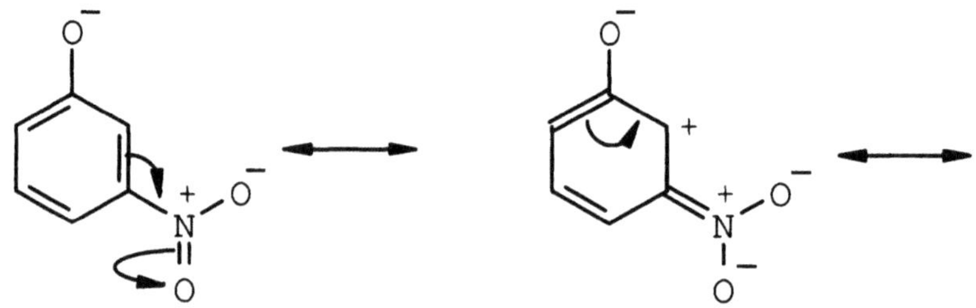

11

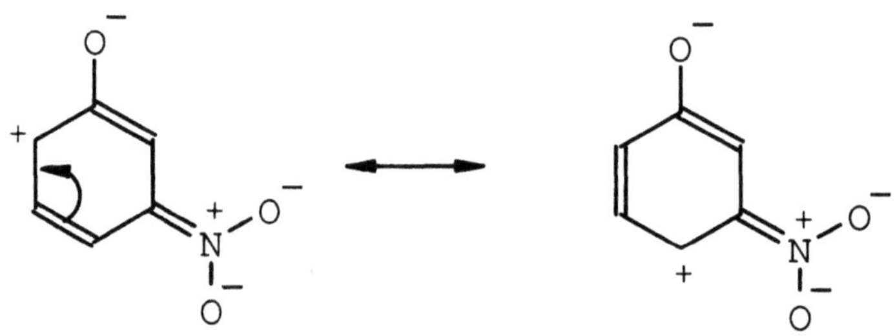

The positions on either side of the ring carbon which bears the $-O^-$ group are slightly positive, and will, by the inductive effect, withdraw electrons from the C–O bond and thus stabilize the negative charge. There is also an electron-withdrawing inductive effect of the nitro group (–I) operating directly on the ring, and this will affect the whole ring and, in turn, be transmitted to the C–O bond. This is also a stabilizing influence. The *m*-nitrophenoxide ion is, therefore, more effectively stabilized than is the phenoxide ion, by the combined electronic effects of the nitro group which finally 'reach' the $C-O^-$ group inductively.

Whilst, in general, if inductive and resonance effects are in opposition the resonance effect usually 'wins', the inductive effect may be operating alone, or with the resonance effect, and must not be forgotten.

? *Arrange the following compounds in order of increasing basicity:*
aniline; p-*nitroaniline;* m-*nitroaniline.*

* *

The correct order, with pK_a *values, is:*

p-*nitroaniline (1.0)* < m-*nitroaniline (2.45)* < *aniline (4.63)*

We are dealing here with a situation very like that of the nitrated phenols. Delocalization in aniline itself, and the consequences of protonation, were dealt with in frame 5. In the case of the *p*-nitroaniline, the amino lone pair is delocalized right out on to the nitro group oxygen, and the consequences of protonation are rather greater than for aniline:

12

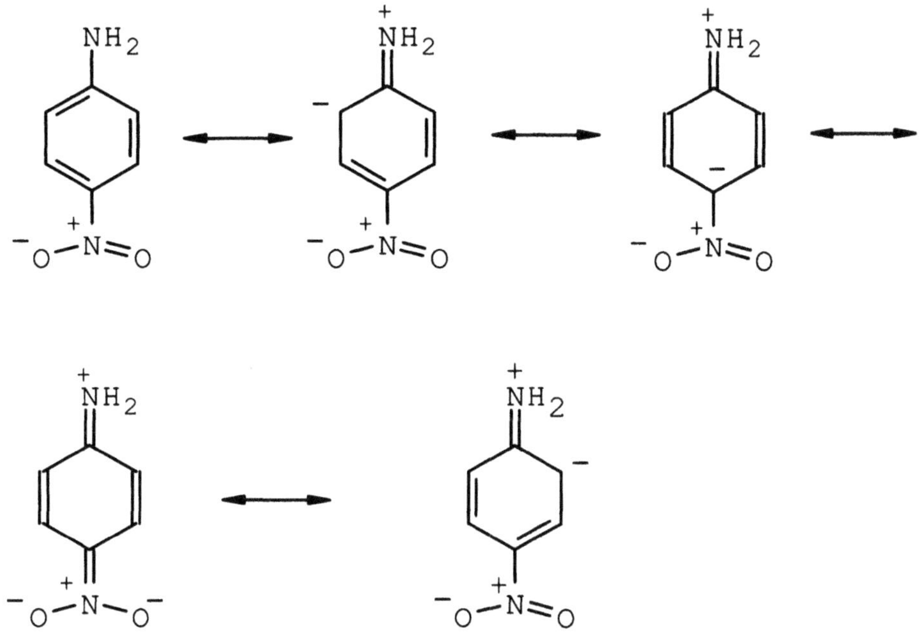

In the case of *m*-nitroaniline, as in the case of the *m*-nitrophenoxide ion, the lone pair cannot be delocalized out on to a nitro group oxygen, but the nitro group 'holds on' to the lone pair through its inductive (–I) effect.

? *Draw out resonance forms to show the interaction between the nitro group and the ring in* m-*nitroaniline.*

* *

You should have drawn out resonance forms as follows:

13

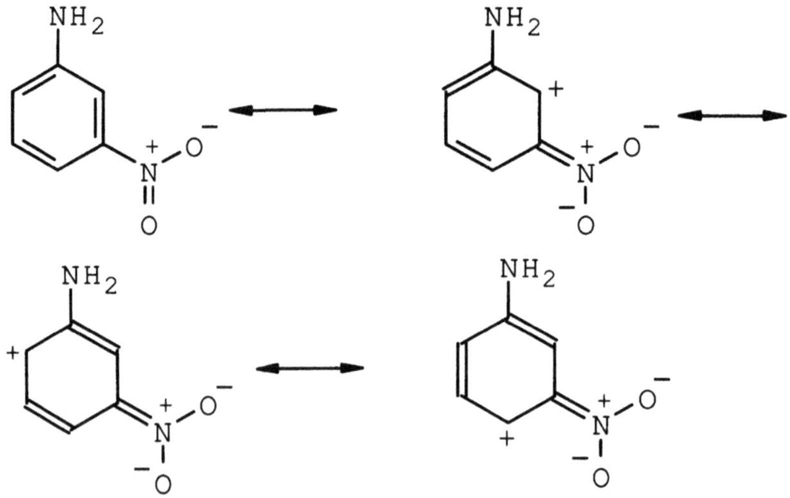

Groups which can only exert an inductive effect will transmit that effect across a benzene ring from virtually any position. We could, thus, easily predict that both *m*- and *p*-trifluoromethylphenol would be stronger acids than phenol itself. It would be harder to predict which of the two trifluoromethyl compounds would be the stronger acid, though.

? *Which would you expect be the stronger base, aniline or* p-*trichloromethylaniline?*

Aniline should be the stronger base, since the trichloromethyl group, with three electronegative chlorine atoms attached to carbon, will have a strong −I effect, and this will destabilize a positively charged protonated amino group on the trichloromethyl compound. Putting it rather more loosely, it will hold on to the lone pair on nitrogen and make it less readily available for binding a proton.

14

Whilst an inductive effect can be transmitted across the π-system of a benzene ring, it is much less easily transmitted along a σ-bonded system. Thus, trifluoroacetic acid is a very strong acid, but when the trifluoromethyl group is moved further from the reactive site, with only σ-bonds between the two, the acidity sharply drops off. This is best illustrated by reference to the sequence:

acid	CF_3CO_2H	$CF_3CH_2CO_2H$	$CF_3CH_2CH_2CO_2H$	CH_3CO_2H
pK_a	0.23	3.07	4.18	4.75

? *Arrange the following group of acids in the order in which you would predict their acidities to increase.*
FCH_2CO_2H; $ClCH_2CO_2H$; $BrCH_2CO_2H$; CH_3CO_2H

On the basis of the argument so far, the prediction should be that the acidities should increase with the increasing values of the electronegativities of the halogens:

The observed pK_a values are:

acid	CH_3CO_2H <	$BrCH_2CO_2H$ <	$ClCH_2H$ <	FCH_2CO_2H
pK_a	4.75	2.90	2.86	2.59

Although these figures tie up with the predictions based upon electronic effects, there is a danger. Firstly, the differences between them are very small – too small for us to feel
15 able to predict them with confidence, particularly given the nature of our initial assumptions (frame 7). Secondly, experimental evidence indicates that the most important factor operating here is ΔS, and not ΔH, and that it is merely fortuitous that the predicted order matches experimental measurements. Indeed, if one were to measure the pK_a values at, say, 80°C instead of 25°C, then one would arrive at a different sequence. So, whilst it turns out that we get the 'right' answer at room temperature by applying the simple electronic rules, we are reminded that they are more in the nature of rules of thumb than of watertight, rigorous theoretical rules.

? *Arrange the following compounds in the order in which you would expect their acidities to increase: $ClCH_2CO_2H$; Cl_2CHCO_2H; Cl_3CCO_2H*

* *

The expected sequence is:

$ClCH_2CO_2H < Cl_2CHCO_2H < Cl_3CCO_2H$

The observed pK_a values are:

2.86 1.26 0.64

Differences are larger here than in frame 15, but the reservations must still stand.

So far we have dealt with acids which have their available protons attached to the usual electronegative elements such as oxygen, and bases which accept protons on to the usual electronegative elements containing lone pairs, such as nitrogen. Sometimes, however, by
16 stabilizing a carbanion, resonance can confer acidity upon C–H bonds.

Ethyl phenyl ketone (1-phenyl-1-propanone) has a relatively acidic α-C–H group, for example:

$$C_6H_5{-}CO{-}CH_2{-}CH_3 + B \longrightarrow C_6H_5{-}CO{-}\overset{-}{C}H{-}CH_3 + BH^+$$

('α' here means adjacent to the carbonyl group –CO–C–C–C etc.
α β γ etc.)

Carbon, with no lone pair, may also be basic if a suitable pair of electrons is available for binding a proton. Alkenes, containing π-electrons, for example, are basic:

$$CH_2{=}CH_2 + H^+ \longrightarrow CH_3{-}\overset{+}{C}H_2$$

? *Write resonance forms for the carbanion derived from ethyl phenyl ketone by removal of the α-hydrogen as a proton.*

* *

You should have written the forms:

It is worth noting here that an α-C–H is situated relative to the carbonyl group in the same way as is the O–H of a carboxylic acid. The acidity of both is enhanced by delocalization of the negative charge in the anion (and the associated lone pair) into the carbonyl group.

Compare:

17

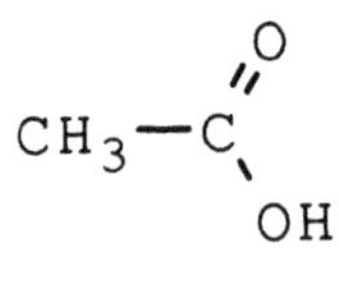

with

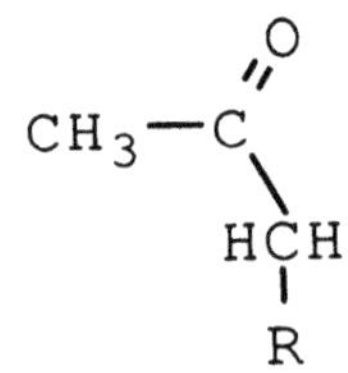

The α-C–H group in esters is also rendered acidic in the same way by the carbonyl group; the effect is not confined to ketones and aldehydes.

? *Write resonance forms for the anion derived by removal of the α-hydrogen from ethyl acetate.*

* *

The appropriate forms are:

It is important to remember at this stage that there are other resonance forms which one can write for the ethyl acetate anion, and these are analogous to the forms that one

18 would have written for ethyl acetate itself:

The lone pair on the alkoxy oxygen is also delocalized into the carbonyl group, and in this sense it is in competition with the carbanion lone pair. Stabilization of the anionic centre is therefore less effective in the ester anion than in the ketone anion.

? *Arrange the following molecules in the order in which you would expect their acidities at the methylene group (CH_2) to decrease.*
$CH_3COCH_2COCH_3$; $CH_3COCH_2CO_2C_2H_5$; $CH_2(CO_2C_2H_5)_2$.

* *

You should have written:

$CH_3COCH_2COCH_3 > CH_3COCH_2CO_2C_2H_5 > CH_2(CO_2C_2H_5)_2$

Approximate pK_a values are:

8.8	11	13.5

19 (acetone has a value of about 20)

All three methylenes are 'activated' by two carbonyl groups, so that the anion, derived by removing one of the methylene hydrogens as a proton, is stabilized by delocalization of the negative charge into two carbonyl groups. However, as we have seen above, the ester carbonyls are less effective at this delocalization than are ketone carbonyls, hence the more ester carbonyls involved, and the fewer ketone carbonyls, the less acidic is the methylene group.

? *Write resonance forms for the anion derived from acetylacetone, $CH_3COCH_2COCH_3$.*

* *

The correct resonance forms are:

20

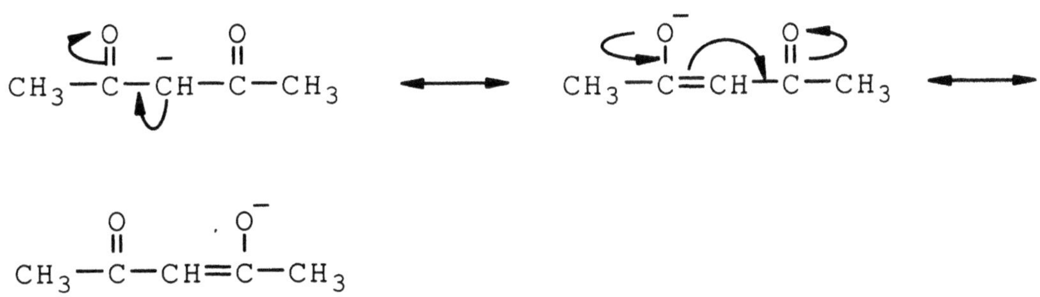

Carbonyl groups are not the only groups which can stabilize an adjacent negative charge. Any electron withdrawing group, particularly if (like carbonyl) it can conjugate with the charge/lone pair, will be able to do this. Typical examples of such groups would be C≡N; P=O; N=O.

? *Account for the fact that nitromethane, when treated with sodium methoxide, a strong base, yields a sodium salt, $Na(CH_2NO_2)$.*

* *

The methyl hydrogens of nitromethane are rendered acidic by the effect of the nitro group which stabilizes the anion derived by their loss:

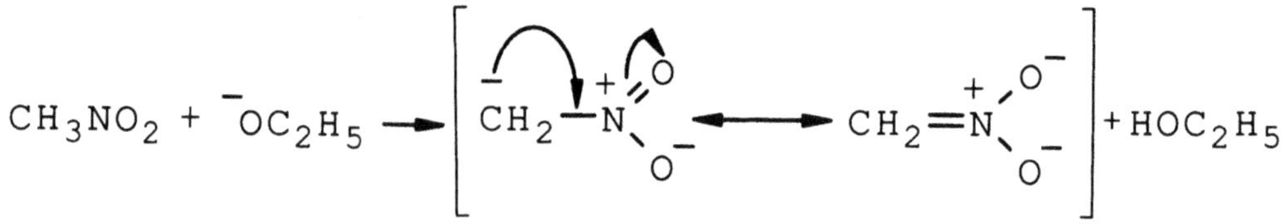

Nitromethane has a pK_a of 10.2 and forms salts with strong bases.

21 One limitation to the approach which we have so far taken to resonance is that it does not make clear that there are certain geometrical criteria which must be met before resonance is a possibility. In order to understand this, it is perhaps best to consider the molecular π-orbitals, and the atomic *p*-orbitals from which we may construct these, which correspond to the delocalized systems which we have been discussing. Thus, if there is to be delocalization of π-electrons across a series of σ-bonded atoms, then the *p*-orbitals on each atom must be parallel for maximum overlap leading to the most stable delocalized molecular π-orbitals. The following diagrams of the acetate ion and 2-propenal (a conjugated system) illustrate the point:

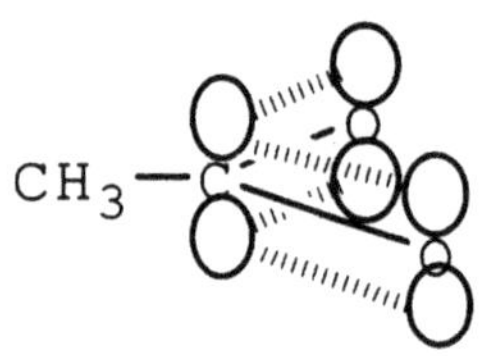

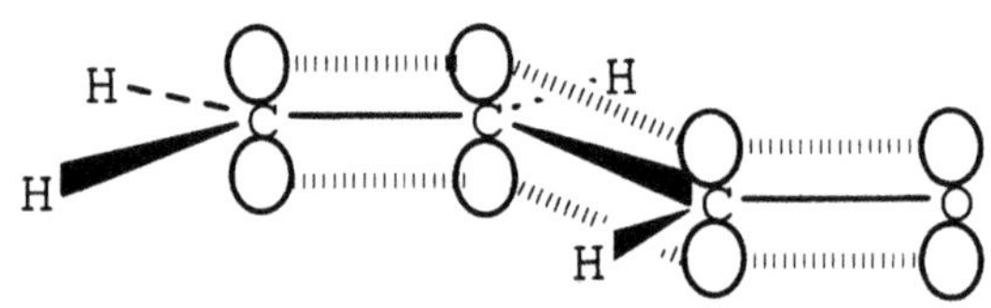

A consequence of this requirement for the *p*-orbitals to be parallel is that the systems must be flat (or at any rate nearly so). Clearly, a three atom system such as $-CO_2^-$ can only be flat. However, the conjugated molecule, 2-propenal, does have an option. Rotation about the bond between C1 and C2 would result in the aldehyde and the alkenyl groups lying in different planes, and this would reduce or prevent delocalization and its consequent stabilizing effect. To achieve maximum stability 2-propenal must be flat, i.e. the aldehyde and alkenyl groups must be coplanar.

? *Given what you have just read about geometrical restrictions on resonance, predict the geometry of nitrobenzene.*

* *

Nitrobenzene, in order to achieve maximum delocalization between the ring and the nitro group, must be flat, with the ring and the nitro group coplanar:

22

It is a good idea to support such a prediction with a diagram showing the *p*-orbitals on each interacting atom.

? *Sketch the nitrobenzene molecule σ-structure to give a sideways view (as on the previous page) and show on it the interacting atomic* p*-orbitals:*

* *

Your diagram should look similar to this one:

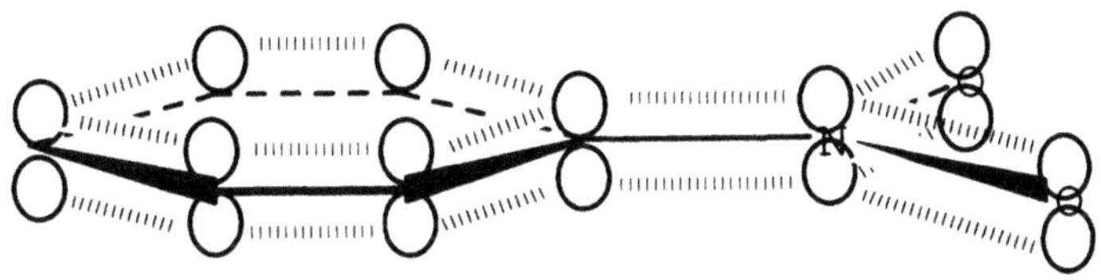

An obvious corollary from this geometrical requirement is that if it cannot be met for some reason, then resonance and stabilization will be prevented or reduced. These consequences show up very clearly in the area of acid and base strengths.

23 ? *Account for the following differences in* pK_a *values:*

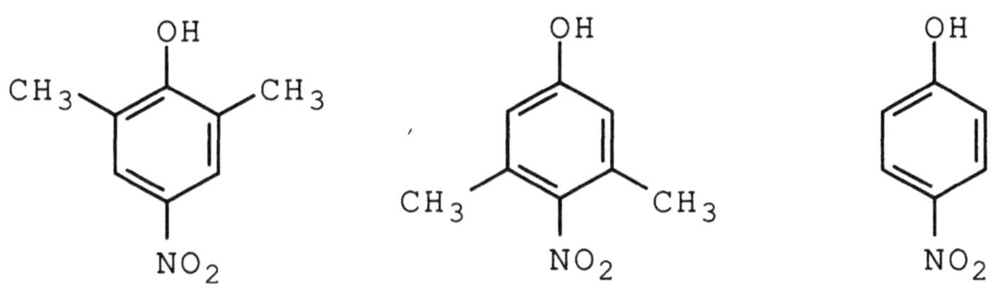

A	B	C
7.16	8.24	7.14

* *

*In compound **C**,* p-*nitrophenol (which we have met before), the OH acidity is enhanced by delocalization of the negative charge in the anion out on to the nitro group. The same is true for compound **A**, which has essentially the same* pK_a *value. However, compound **B** is a substantially weaker acid and this can be attributed to the effect of the two bulky methyl groups close to, and on either side of, the nitro group. Their size, and that of the nitro group, dictate that they are trying to occupy space which the nitro group requires if it is to be coplanar with the ring. The only way to avoid this steric interaction is for the nitro group to twist out of the ring plane, so that its own plane presents a face to each methyl group and it can fit in neatly between them. This prevents resonance stabilization of the anion, and the nitro group can exert a stabilizing influence on the anion only through its inductive effect (–I). **B** is therefore a weaker acid than either **A** or **C**.*

24

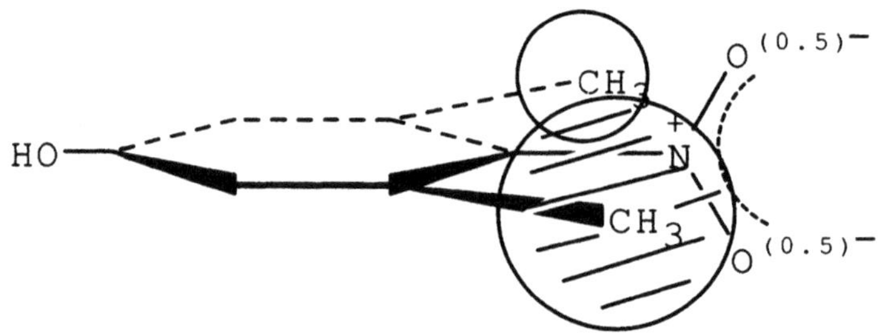

This effect is known as **steric inhibition of resonance** and can be observed in a variety of systems.

Steric interactions of this type are not the only way of suppressing resonance through geometrical restriction. The σ-bonded structure of molecules may also dictate that certain groups which would otherwise interact by a resonance effect cannot do so:

? *Explain why compound **D*** (pK_a = *7.79) is a stronger base than compound **E*** (pK_a = *5.06).*

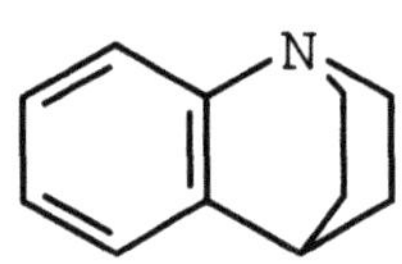

D

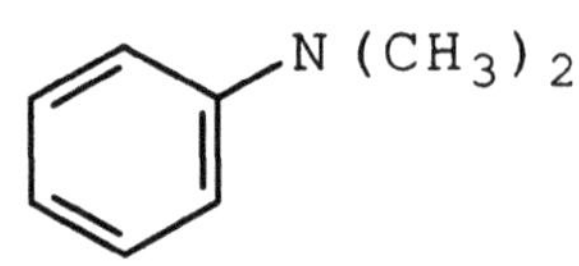

E

* *

*In compound **E**, as in aniline, the nitrogen lone pair is delocalized into the ring and its basic character is reduced. In compound **D**, however, the nitrogen lone pair is held in an* sp^3 *orbital in the same plane as the benzene ring. It is therefore orthogonal (perpendicular) to the ring carbon* p-*orbitals and cannot interact with them. Resonance is thus prevented, and the lone pair in compound **D** is more readily available for protonation.*

It would at this point be of interest to refer back to the cumulated molecules allene and ketene, referred to in Programme 2, frames 14 and 15.

25 A diagram of the π-orbitals of allene is shown below. The molecule is linear, with the two CH_2 groups orthogonal (i.e. lying in planes which are mutually perpendicular). The central carbon atom (C2) is *sp*-hybridized, with its two unhybridized *p*-orbitals

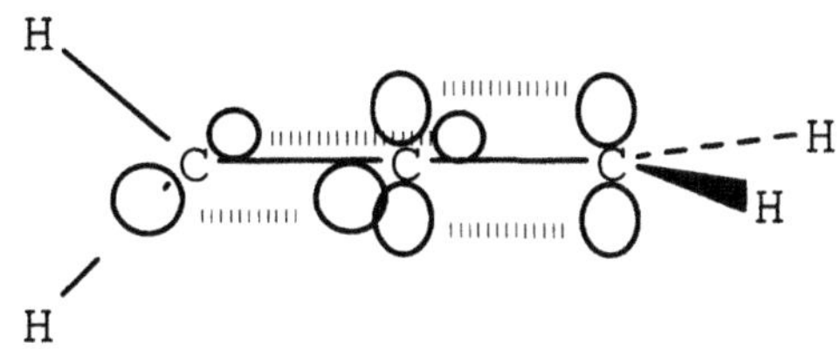

contributing to the π-orbitals of the two double bonds as shown.

? *It is possible now to identify a further (Programme 2) reason why the two double bonds cannot be conjugated. What is this reason?*

* *

26 *The π-orbitals of the two double bonds are mutually orthogonal since the atomic* p-*orbitals on C2 are orthogonal. The two π-orbitals cannot overlap, therefore, and thus cannot interact with each other.*

This is the last frame in the acid/base programme, Programme 3, which you have now completed. You can now proceed to Programme 4 on curly arrows and organic reaction mechanisms.

* *

Programme 4

Curly arrows and organic reaction mechanisms

The primary objective of this programme is to enable you to use curly arrows to describe organic reaction mechanisms. Inevitably this means that you will be looking at mechanisms too, and learning some of their basic features, and you will, in the course of the programme, meet three of the four fundamental reaction types: **substitution**, **elimination** and **addition.** Rearrangement reactions you will meet elsewhere. As you go through the programme, learning and practising how to use curly arrows in this context, you will also work your way through these three reaction types, and they will be summarized at appropriate points.

1

Curly arrows mean exactly the same in mechanism as they mean in resonance; they represent the shift of a pair of electrons. In mechanism, however, instead of moving from one hypothetical species (or canonical form) which has no real existence, to another, we shall actually be moving from one chemical species to another, with electron pairs moving to form new bonds between atoms and to break old ones.

It is just as important to use the arrows accurately here as it is in writing resonance forms. The arrows are written on the reactants, on the left-hand side of an equation, and indicate the electron pair movements which are necessary to generate the products shown on the right-hand side of the equation. The tail of an arrow should be placed where an electron pair starts from, and the head where the pair is going to. Thus, we can represent the formation of a bond between nitrogen and hydrogen, when ammonia is protonated (forms a salt), as follows:

The ammonium ion should strictly be written as $\overset{+}{N}H_4$ (or $H_4\overset{+}{N}$) of course, since all four hydrogens are equivalent. However, it is convenient here to emphasize the newly formed N–H bond. Note that we do not represent the lone pair on the nitrogen; it is taken as read, as you learnt in the resonance programme (Programme 2).

When aqueous sodium hydroxide reacts with an acid, say hydrochloric acid, then the reaction is one in which, essentially, the hydroxide ion is protonated. The sodium and chloride ions do nothing.

$$Na^+ + HO^- + H^+ + Cl^- \longrightarrow Na^+ + Cl^- + H_2O$$

? *Write, using curly arrows, an equation to describe the protonation of the hydroxide ion.*

* *

The equation which you should have written is:

A naked proton, of course, does not exist in aqueous solution, which is the medium in which most of the simple acid/base reactions which you have met will have been carried out. It usually exists as the hydroxonium ion, H_3O^+, and this is further solvated by more water molecules so that we should write it as $H_3O^+_{(aq)}$. In practice we find it convenient to write it simply as H_3O^+. Nor does a naked proton exist in other solvents; in a non-aqueous
2 solvent of low dielectric constant, such as benzene, or in the gas phase, an acidic compound such as HCl, for example, will exist simply as HCl molecules. We would therefore write the protonation reaction of ammonia with HCl, in benzene solution, as:

Note that the chloride ion departs from the hydrogen with its bonding pair to make room round hydrogen for the incoming electron pair from ammonia, the nucleophile (see frames 3–5).

? *Write an equation describing the protonation reaction of hydroxide ion in aqueous solution, using curly arrows and using the hydroxonium ion as a source of the proton.*

* *

The equation which you should have written is:

In the mechanistic context, we refer to the species donating the electron pair to form a bond as a **nucleophile** and to the complementary species accepting the electron pair as an **electrophile.** From the examples which we have chosen here, it will be clear that a nucleophile is a base and an electrophile is an acid.

3 The terms 'nucleophile' and 'base' are not synonymous, however. The former reflects the **kinetic** aspects of electron pair donation, whilst the latter reflects the **thermodynamic** aspects. Thus a powerful nucleophile is one which reacts rapidly with an electrophile (small energy of activation) and a strong base is one whose reaction with an acid reaches an equilibrium position well over to the right (large, negative $\Delta G°$). Furthermore, a nucleophile is defined in terms of its capacity to react with *any* electrophile, whereas, at any rate on the Brönsted definition, a base is defined solely in terms of its ability to accept a proton.

Similarly, the term 'acid' is a thermodynamic term, and 'electrophile' a kinetic term.

? *What are the quantitative terms used to express the strengths of acids and bases, and what kind of terms are they?*

* *

The terms are K_a *or* pK_a. *They are equilibrium constants or negative logarithms (base 10) of equilibrium constants.*

Note that an equilibrium constant says nothing about the rate of a reaction, which is the concern when we are talking about nucleophiles and electrophiles, and that a strong base is, therefore, not necessarily a powerful nucleophile. In practice, the two often do run parallel, but this is not always so.

The nucleophiles which we have met so far are able to donate electron pairs because they have lone pairs. Although this is perhaps the most common feature which will confer nucleophilicity upon a species, it is not the only one. Thus, as we shall see later, π-electrons may also be used as a source of the electron pair to be donated. In spite of what you may
4 have read before, please note that **a negative charge is not in itself able to give rise to nucleophilicity.** There are plenty of examples of negative ions which are unable to donate electron pairs. The boron atom in the BF_4^- ion, for example, carries a formal negative charge, but has no electron pairs which it can donate. There are also plenty of good nucleophiles which carry little or no negative charge. Similarly, **a positive charge is neither necessary nor sufficient to confer electrophilicity upon a species.**

? *Which of the following species would you expect to be nucleophilic? Indicate which atom in each would be the nucleophilic site, and do not worry about whether they would be strong or weak nucleophiles.*

$(CH_3)_3N$,	H_2O,	CH_3^-,	CH_4	$(CH_3)_3P$
(i)	(ii)	(iii)	(iv)	(v)

* *

All but CH_4 *are nucleophiles. The nucleophilic sites, each possessing a lone pair or pairs, are:*

(i) N, (ii) O, (iii) C, (iv) none, (v) P

Given that most of the atoms with which we are concerned in organic chemistry are of first row elements, they cannot hold more than eight electrons in their outer shells. Thus, in order to be an electrophilic site, they must either possess an incomplete outer shell or be able to 'get rid' of a pair of electrons from their full outer shell to make room for a new pair. The electrophiles which we have met so far illustrate both of these categories. Thus,
5 the naked proton has an empty outer shell; it is formally **electron deficient.** It can hold two electrons in its outer shell, and is therefore able to accept a pair of electrons from a nucleophile. The hydroxonium ion and hydrogen chloride, however, have to make room round their (potential) proton by losing either a water molecule or a chloride ion:

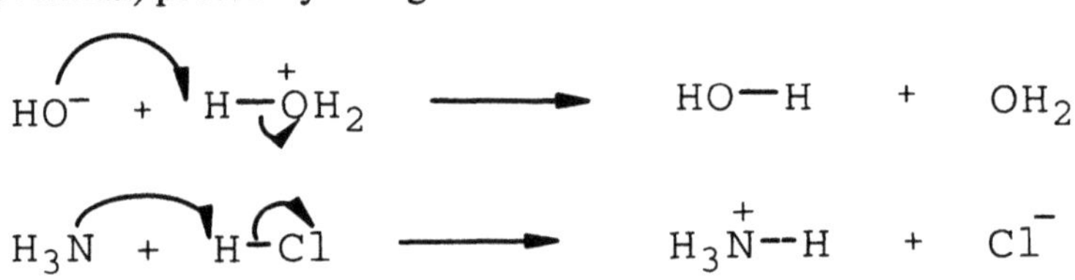

In neither case is the reacting hydrogen atom formally electron deficient.

The group which departs in this way, with its bonding pair of electrons, is known as a **leaving group** (or negative leaving group). A useful guide as to how good a leaving group will be is how weak it is as a base or, more conveniently (but the same thing really), the strength of its conjugate acid. H_2O and Cl^-, for example, are good leaving groups because they are weak bases (i.e. their conjugate acids, H_3O^+ and HCl, are strong).

? *Which of the following species would make good leaving groups?*

OH^-, F^-, I^-, NH_2^-, $^-O–PO_3H_2$ (phosphate anion)

* *

The good leaving groups here are I^- and $^-O–PO_3H_2$ since their conjugate acids (HI and phosphoric acid) are strong. The conjugate acids of all of the other species are weaker: H_2O, HF, NH_3.

6 As well as getting rid of a leaving group, there are other ways in which room can be made at an electrophilic site which is not electron deficient for a pair of electrons coming in from a nucleophile. One in particular which we shall mention here is the movement of a π-electron pair in a multiple bond. Thus, the carbonyl group is electrophilic at carbon because it can shift the π-electron pair over on to the oxygen atom. Since the oxygen atom is well able to bear the resulting negative charge, this is a perfectly feasible process:

? *Which of the following groups containing multiple bonds would you expect to be electrophilic? Specify the electrophilic site, and give the reasons for your choice.*

* *

The cyanide (nitrile) group is electrophilic at carbon; the alkenyl system, however, is not electrophilic. If a nucleophile attacked one of the alkene carbon atoms then the other would have to bear a negative charge. Carbon (except in certain rather special circumstances – Programme 3 on acids and bases) is generally unwilling to bear a negative charge, whereas nitrogen is perfectly capable of doing so.

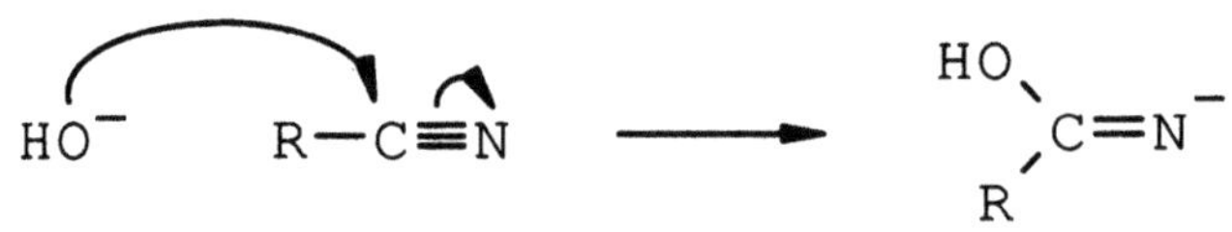

7

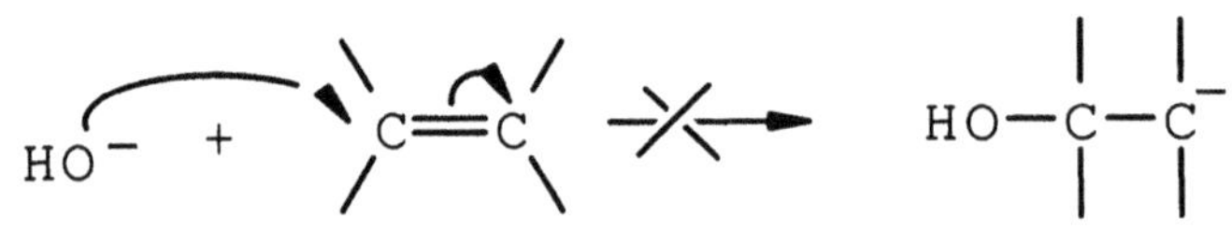

(not allowed)

Summarizing, we can say that **a nucleophile is an electron pair donor**, and **an electrophile is an electron pair acceptor**. A lone pair is the most common structural feature which can confer nucleophilicity upon a molecular site (π-electrons will also do this), and electron deficiency or the ability to make room for an incoming electron pair will confer electrophilicity. Attachment to a good leaving group (which must have a strong conjugate acid), or involvement in a multiple bond to an electronegative atom, are the common features which enable sites to make room for an incoming electron pair.

? *Classify the following species as electrophiles or nucleophiles, and give the reasoning behind your choice.*

CN^-, $B(CH_3)_3$, $C_2H_5O^-$, C_2H_5OH, CH_3I, $CH_3C{\equiv}CCH_3$

* *

The correct classifications are as follows:

CN^- *is a nucleophile with a lone pair on each atom.*
$B(CH_3)_3$ *is an electrophile since the boron atom is electron deficient.*
$C_2H_5O^-$ *is a nucleophile since the oxygen carries three lone pairs.*
C_2H_5OH *is a nucleophile since the oxygen carries two lone pairs.*
CH_3I *is an electrophile at carbon since a good leaving group (I) is attached there.*
$CH_3C{\equiv}CCH_3$ *cannot be an electrophile since a carbon would have to bear a negative charge to allow this (frame 7). With π-electrons, however, it can be a nucleophile.*

Return now to the example from frame 2:

8

Consider this reaction from the point of view of the HCl molecule. The chlorine atom has been replaced by an $H_3\overset{+}{N}$ group, i.e. the latter group has been substituted, in the HCl molecule, for the former. In short, this is a substitution reaction, albeit of a very simple sort. One could say that, because the chloride ion is a good leaving group, when the ammonia molecule hits the hydrogen atom of HCl, provided it is oriented so as to present its lone pair to the hydrogen atom, then it can knock the chloride ion off. Such collisions are inevitably going to be frequent in a mixture of the two compounds, either in the gas phase or dissolved in benzene, and many of these will be energetic enough to bring the reaction about.

? *Using the above reaction as a model, and with the help of curly arrows, write an equation to describe the outcome of a collision between an ammonia molecule and methyl iodide.*

* *

The equation which you should have written is:

We are now firmly into organic chemistry, and we classify this as a **nucleophilic substitution**. It is an S_N reaction. As organic chemists, we 'sit' on the organic molecule and decide that we have been attacked by a nucleophile, hence we call it *nucleophilic* substitution. This is arbitrary, of course, since it is equally true that the ammonia has been attacked by an electrophile. However, if we always classify reactions by deciding what attacks the organic reagent, then we have a consistent classification of reactions, and everyone uses the same terms to mean the same things. If two organic compounds react together, then we have to decide which one to sit on to classify the reaction. In practice this is less of a problem than it seems; consensus as to what to do here turns out to be readily available and it does not not lead to inconsistency. We shall see how this works out later in the programme.

9 There are, in fact, two ways in which this simple type of reaction might occur. It is possible that the leaving group might go before the nucleophile hits the alkyl halide, or it might all go in one step as above. The two step alternative can be illustrated by taking another example, in this case the reaction between tertiary butyl bromide and water.

Step 1: $(CH_3)_3C\text{—}Br \longrightarrow (CH_3)_3C^+ + Br^-$

Step 2: $(CH_3)_3C^+ + H_2O \longrightarrow (CH_3)_3C\text{—}\overset{+}{O}H_2$

$$(CH_3)_3C\text{—}\overset{+}{O}H_2 + H_2O \longrightarrow (CH_3)_3C\text{—}OH + H_3O^+$$

The final equation is merely the rather rapid transfer of a proton from one basic site to another, and is not really a part of the substitution process itself. It does, however, show that the product of the reaction is an alcohol.

? *Rewrite the equations for steps 1 and 2, inserting into them the appropriate curly arrows.*

* *

The equations which you should have written are:

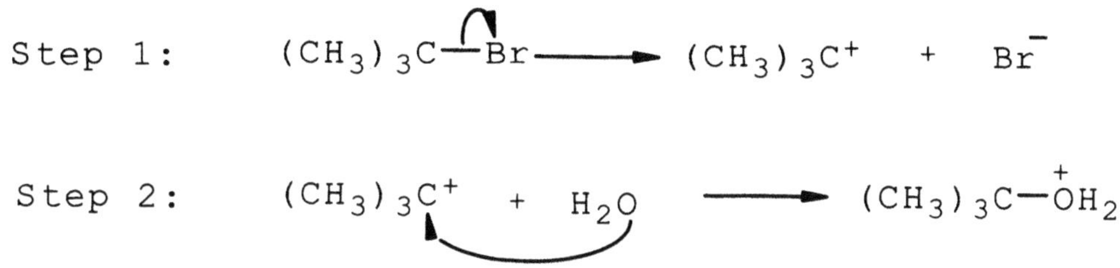

10 For reasons which need not be of present concern, the one step process is called an S_N2 reaction, and the two step process an S_N1 reaction. (The numbers used refer to the number of reactant molecules involved in the transition states.)

A third option for the nucleophilic substitution reaction is for the nucleophile to attack first, to form an intermediate compound (or ion), and for the leaving group to depart in a second step. In practice this is not observed for substitution at saturated carbon.

? *Attempt to write equations, with curly arrows, to describe this third option for the reaction between methyl iodide and a hydroxide ion. Suggest a reason to explain why this reaction pathway is not observed.*

* *

The equations which you might have written are:

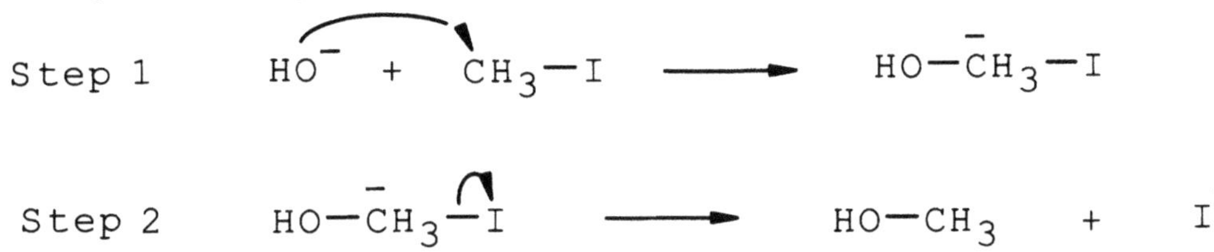

However, since step 1 involves the production of an intermediate anion with five covalent bonds to carbon, i.e. in which the 'octet rule' has been broken, it is apparent that this step cannot take place.

11 If a saturated organic compound contains a good leaving group, and a hydrogen atom two carbons away from the leaving group, then it is possible for an electron pair donor to attack such a molecule in a rather different way. The donor may behave as a base and remove a proton and the result is an **E** or a **1,2-elimination** reaction. The figures merely indicate that the groups eliminated are on adjacent carbon atoms:

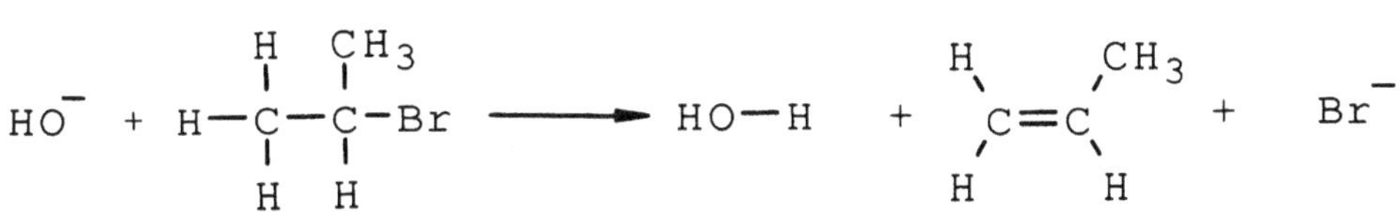

? *Rewrite this equation complete with curly arrows. Think back and recall how curly arrows can be linked and used to shift an electron pair from one bond to an adjacent one (Programme 2).*

* *

The equation which you should have written is:

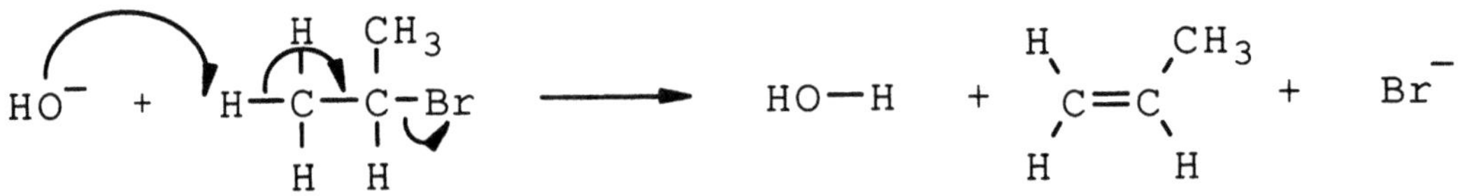

Note how the C–H electron pair moves from bond to adjacent bond (Programme 2, frame 5)

Like the S_N reactions, these elimination reactions can proceed in more than one way, reflecting the different possibilities for timing the making and breaking of the various bonds involved (frame 9). Like the S_N1 and S_N2 reactions for example, we have E1 and
12 E2 reactions. The equations above describe the E2 process; the E1 process is as follows:

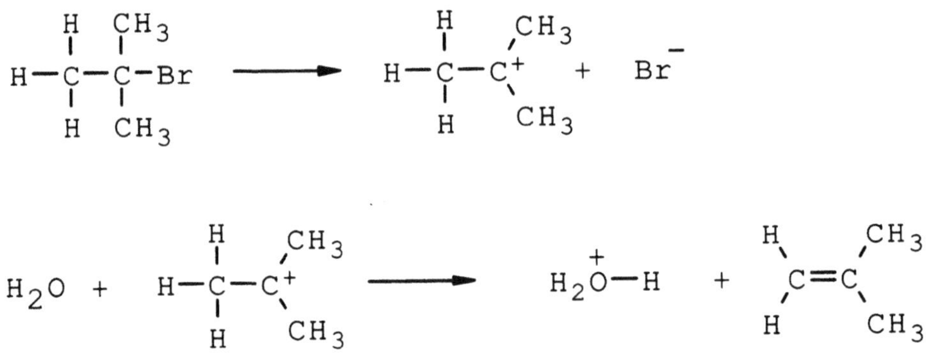

? *Rewrite the above equations, complete with curly arrows.*

* *

The equations which you should have written are:

13

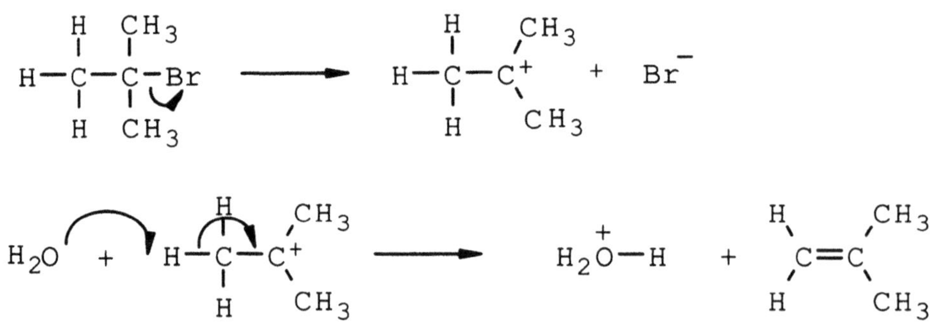

We have now looked at several ways in which a saturated organic compound containing a leaving group may react with electron pair donors, and this has involved nucleophilic substitution (S_N1 and S_N2) and elimination (E1 and E2) reactions.

When a nucleophile reacts with an unsaturated centre, like the carbonyl group of aldehydes or ketones, then an addition reaction commonly ensues. Thus, when a cyanide ion attacks the carbonyl group of acetaldehyde, addition takes place. The anion which is first formed may be protonated by undissociated hydrogen cyanide, a weak acid, provided that the solution is not too alkaline:

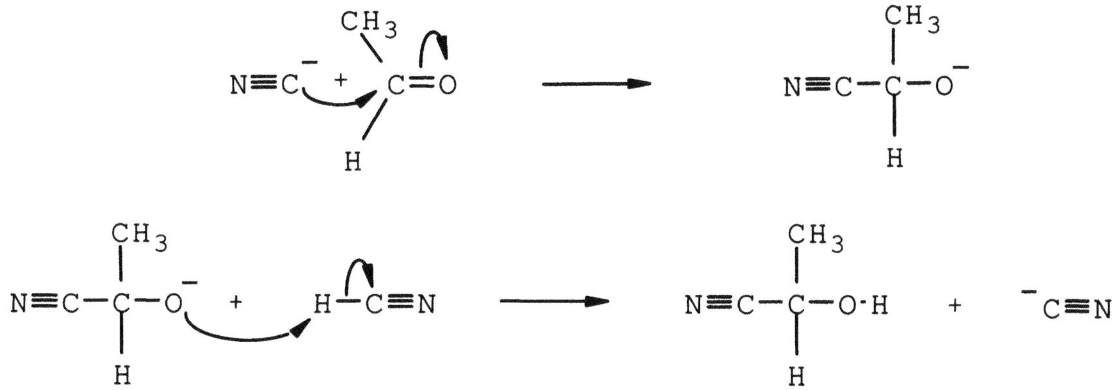

Sitting on the organic compound again, we recognize that it has been attacked by a nucleophile and we therefore call this type of reaction **nucleophilic addition.**

The nucleophilic addition of hydrogen cyanide to an aldehyde or ketone is unusual in that it is catalysed only by bases; it will not take place if the solution is acidic, nor even if it is too alkaline. Many other additions, however, are catalysed by both acids and bases.
13 Water, for instance, will add to aldehydes and ketones in a reaction which may be acid- or base-catalysed. The base-catalysed version is very similar to the addition of hydrogen cyanide; the acid-catalysed version may be written as follows:

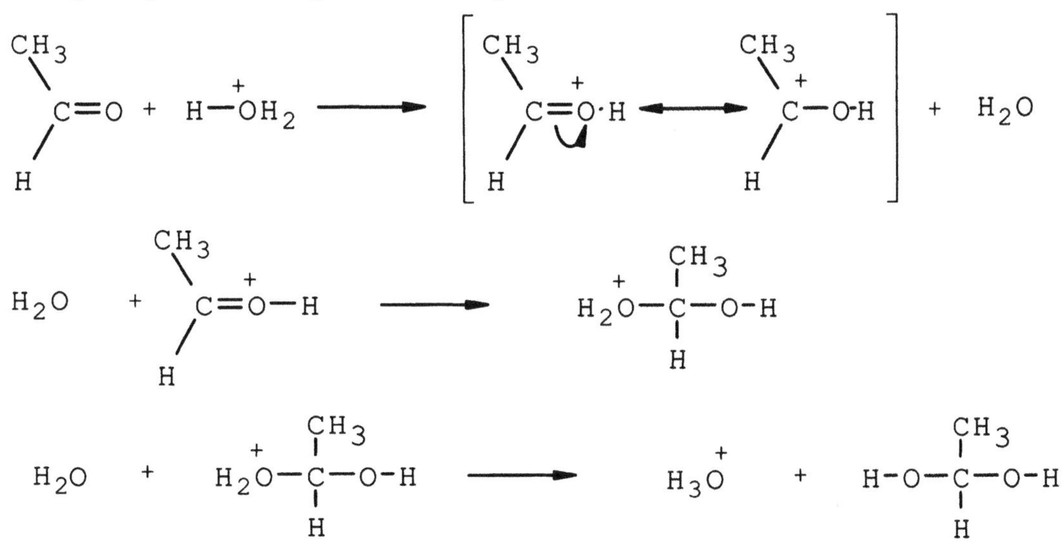

Note that the cation which is formed initially is stabilized by delocalization.

? *Rewrite the above equations for the acid-catalysed hydration of acetaldehyde and incorporate the appropriate curly arrows.*

* *

The equations which you should have written are:

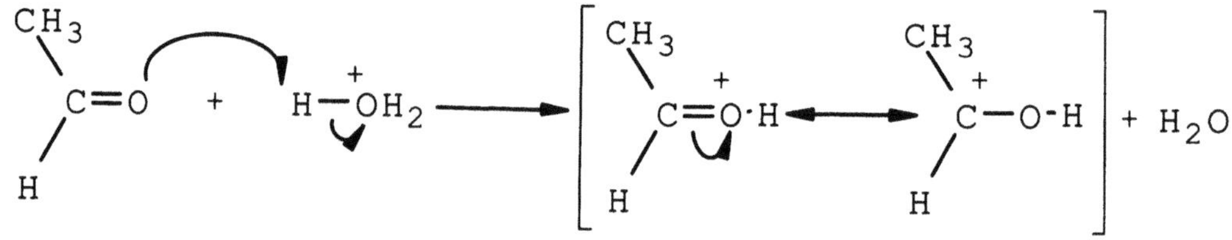

14

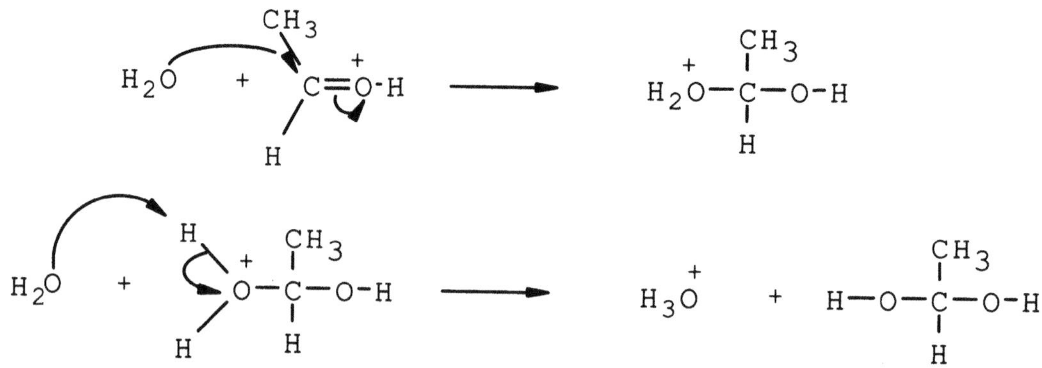

Exactly parallel to the formation of a hydrate, by addition of water, is the reaction of aldehydes (and ketones) with alcohols to give a hemi-acetal (or hemi-ketal).

? *Write equations with curly arrows to describe the formation of the hemi-acetal by the acid-catalysed addition of methanol to acetaldehyde.*

* *

The equations which you should have written are:

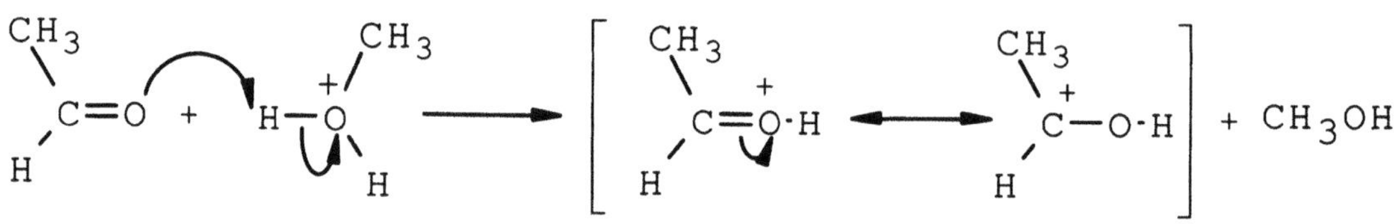

15

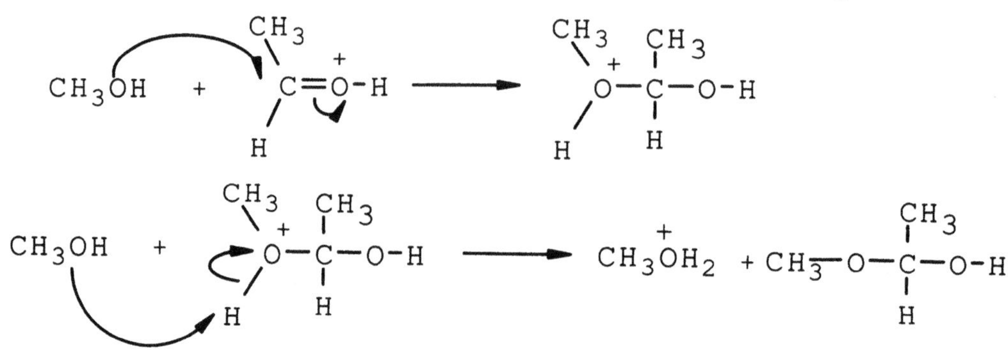

In the presence of acid, further protonation of basic sites in the hemi-acetal may occur. If one of the hydroxyl oxygens is protonated, then it becomes a good leaving group and may dissociate from the rest of the molecule. This is assisted considerably by the resonance stabilization of the resultant cation:

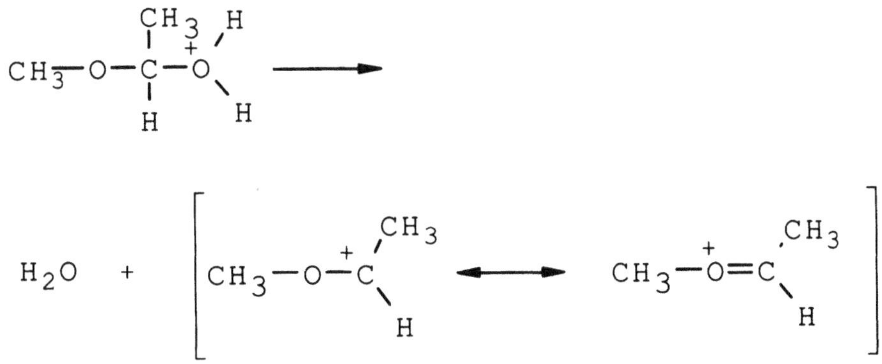

If an excess of methanol had been used initially, then this cation, which is very like the protonated aldehyde, will be attacked by the nucleophilic methanol and the product, after removal of a proton, is an acetal:

15

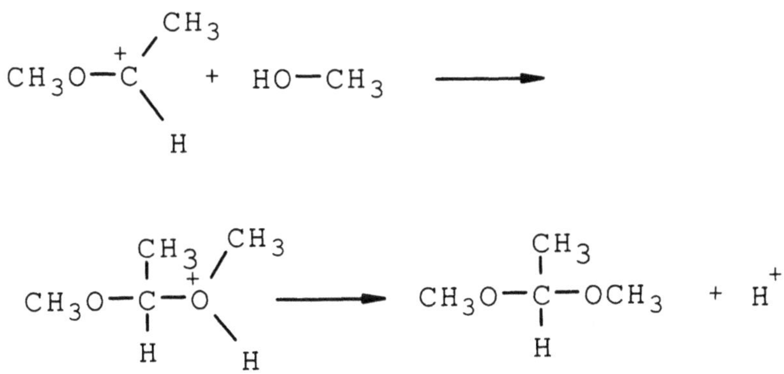

The proton would be removed by a solvent (methanol) molecule, of course, just as it was in the final step of the hemi-acetal production, above. However, to save time and space it is frequently written as though it had just fallen off, as a naked proton, and the role of the solvent is taken for granted. We shall adopt this convenient abbreviation in a number of places in this programme, and you will meet it frequently in textbooks and papers.

If an aldehyde is treated with acid and an excess of methanol under anhydrous conditions, therefore, it is converted to an acetal. The absence of water is critical, since the equilibrium is nicely balanced. If, on the other hand, the acetal is treated with **aqueous** acid, since the reactions are all reversible it is readily converted back to the aldehyde and the alcohol.

? *Rewrite the above sequence of equations, leading to the acetal, with the appropriate curly arrows inserted.*

* *

The equations which you should have written are:

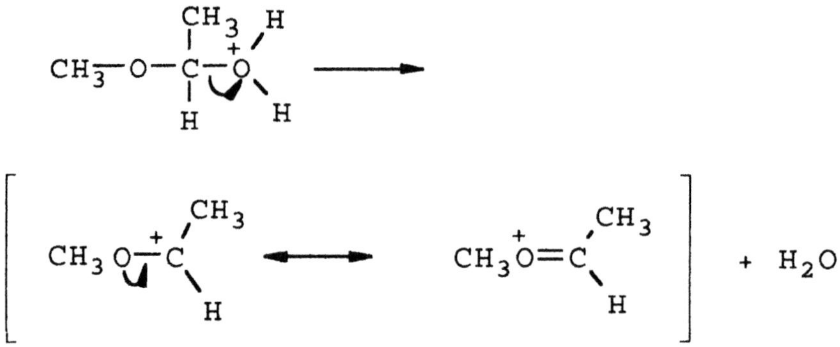

16

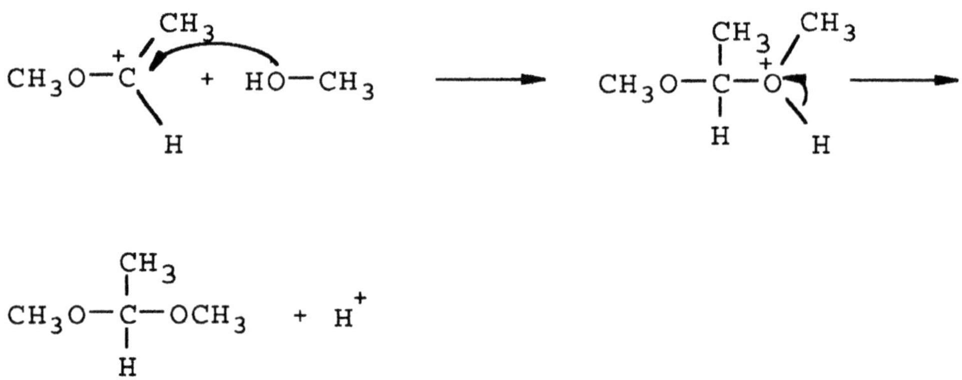

The carbonyl group is not found only in aldehydes and ketones; it is also an essential feature of carboxylic acids and their derivatives such as esters, amides, acid anhydrides and acid chlorides.

? *Using an equation and curly arrows, predict the initial reaction between a hydroxide ion and acetyl chloride (CH_3COCl). Just one step is required here.*

* *

The equation which you should have written is:

Assuming that the reaction is carried out in aqueous solution, where plenty of protons are available from the water, one might expect that the resultant anion here would be protonated to give an adduct, as is the case with aldehydes and ketones. However, in this case, as with all derivatives of carboxylic acids, there is a good negative leaving group attached to the carbonyl carbon. Instead of picking up a proton to form a neutral molecule, it is also possible for this anion to expel the leaving group:

17

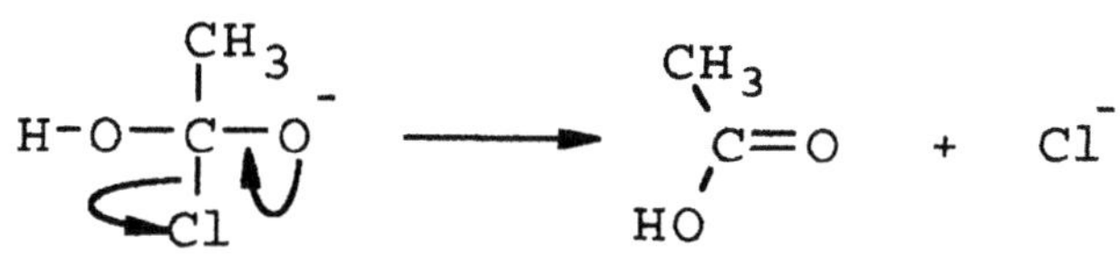

In the presence of hydroxide ions, of course, the acetic acid produced would go on to give the salt, but the essential feature of this whole reaction sequence is that the chlorine has been replaced by a hydroxy group. The overall effect, in other words, is substitution, but it is not a direct substitution like an S_N reaction; it is made up of an addition followed by an elimination step and this type of reaction, characteristic of carboxylic acid derivatives, is sometimes known as an **addition/elimination reaction.** The intermediate adduct is referred to as the **tetrahedral intermediate.**

? *By means of equations and curly arrows, write out the complete reaction pathway for the reaction between aqueous sodium hydroxide and (i) ethyl acetate (ethanoate), (ii) acetic (ethanoic) anhydride.*

* *

The equations which you should have written are:

(i)

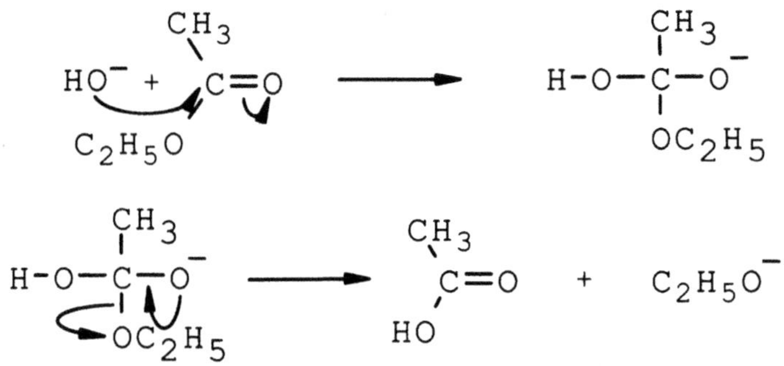

$$CH_3CO_2H \quad + \quad C_2H_5O^- \longrightarrow CH_3CO_2^- \quad + \quad C_2H_5OH$$

(ii)

18

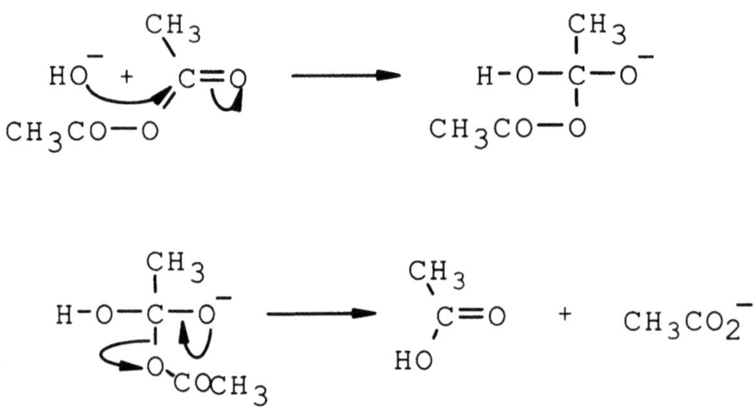

In pretty well all of the reactions which we have covered so far, we have been dealing with processes initiated by the attack of a nucleophile, or electron pair donor, on an organic molecule. We have seen how the organic molecule has accommodated the new electron pair either by expelling a leaving group or by moving π-electrons along on to a suitable adjacent site. This has introduced:

I. nucleophilic substitution at saturated carbon (S_N1 and S_N2);
II. nucleophilic substitution at the carbonyl group of carboxylic acid derivatives (addition/elimination);
III. 1,2-elimination reactions (E1 and E2); and
IV. nucleophilic addition (to aldehydes and ketones).

What we have not yet considered is what happens if an organic molecule is attacked by an electrophile. Such an attack can only be successful in forming new bonds, of course, if the organic molecule can provide an electron pair for the electrophile to accept. In other words, the organic molecule must be a nucleophile.

An organic molecule can be a nucleophile if it contains a functional group with a lone pair, such as the hydroxyl group of alcohols (for example, frames 15 and 16) or the amino group ($-NH_2$) of amines. In such cases the organic molecules will behave like their simpler inorganic relatives, water and ammonia. An organic molecule may also be a nucleophile, however, with carbon as the nucleophilic site, if it contains a π-bond. A carbanion (an anion with the negative charge and a lone pair on carbon) will also have a nucleophilic site based upon carbon; the simplest example of this would be the cyanide ion (see frame 13).

? *How would you expect (i) triethylamine $(C_2H_5)_3N$ (ii) ethylene $CH_2{=}CH_2$ to react with hydrogen bromide?*

* *

Both organic molecules would be protonated:

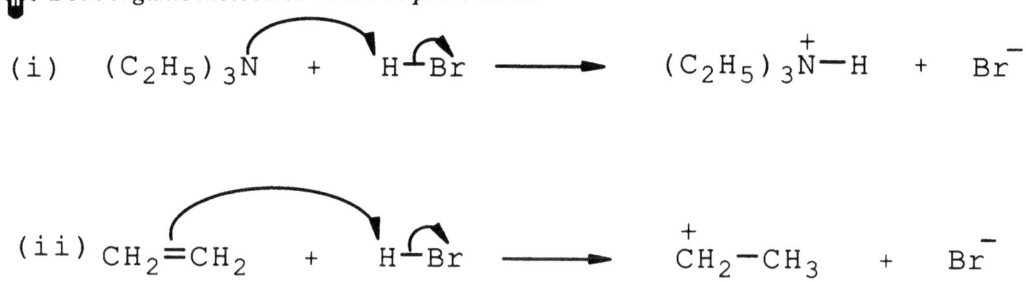

19

In reaction (i) a stable, substituted ammonium salt is produced – triethylammonium bromide. In reaction (ii), however, the product is a **carbocation.** It is electron deficient, having only six electrons round carbon, and is a powerful electrophile. The bromide ion, the other product in this reaction, is a nucleophile and the two react to produce the uncharged, fully saturated, ethyl bromide (bromoethane):

The overall result is the addition of hydrogen bromide to the alkene:

$$CH_2{=}CH_2 \quad + \quad H{-}Br \longrightarrow CH_3{-}CH_2{\cdot}Br$$

Since the reaction was initiated by the attack of an electrophile on the alkene it is classified as an **electrophilic addition.**

? *Predict the effect of (i) hydroxide ion, and (ii) ethylene, upon molecular bromine.*

* *

Since the bromide ion is a good leaving group, we can expect both of these nucleophiles to displace it from a bromine molecule:

(i)

(ii)

20

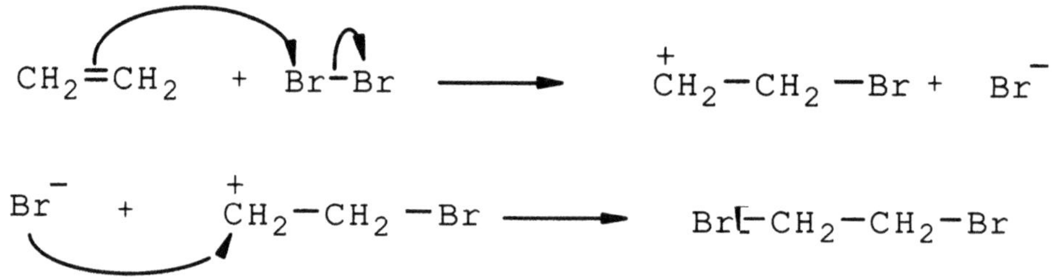

In reaction (ii), as in reaction (ii) of the previous frame, the bromide ion produced reacts with the carbocation to give the fully saturated adduct, and we have another example of electrophilic addition to an alkene.

? *How would you expect benzene to interact with bromine? Only one step is required at this point:*

* *

Note that the carbocation here is stabilized by resonance:

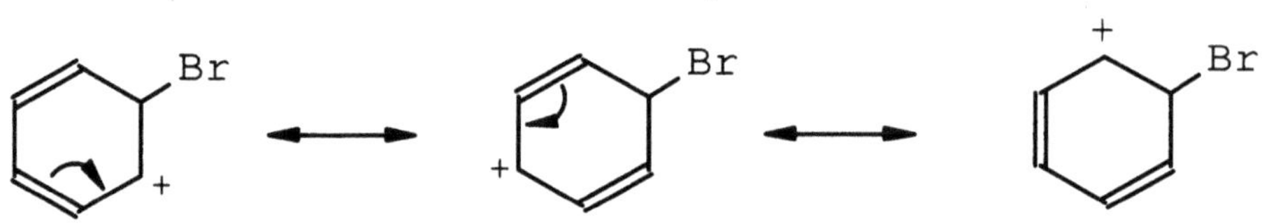

21

This time, if the bromide ion added to the carbocation it would produce a molecule which was not aromatic:

Rather than generate a non-aromatic product like this, it turns out that the bromide ion abstracts a proton from the carbocation to generate a much more stable, aromatic compound.

? *Write down the reaction, complete with curly arrows, for the abstraction of the proton from the carbocation by the bromide ion.*

* *

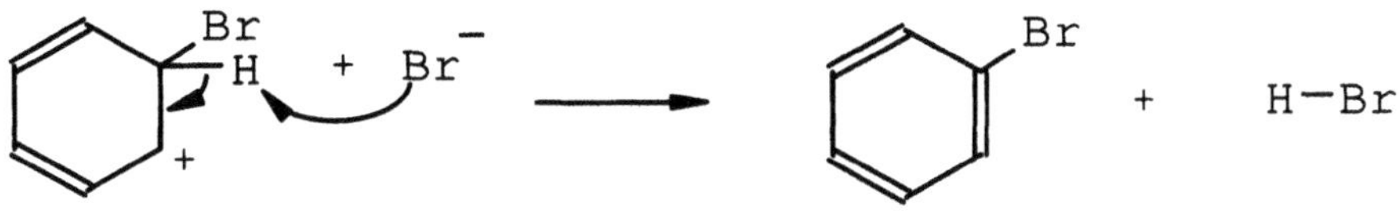

Since there is no water present, the hydrogen bromide will remain unionized.

The net result of this process is a substitution reaction, in which a hydrogen atom of the benzene has been replaced by a bromine atom. This is **electrophilic substitution,** the most characteristic reaction of aromatic compounds.

22

We have now met both types of substitution and both types of addition (nucleophilic and electrophilic in each case), as well as elimination reactions. Although it will not yet be possible to predict from scratch the outcome of more complex reactions, you should be able to write down their products if the reactants, complete with curly arrows, are given, and you should be able to write curly arrows on the reactants if both reactants and products are given. Some exercises of this type follow.

? *What will be the products of the following reaction?*

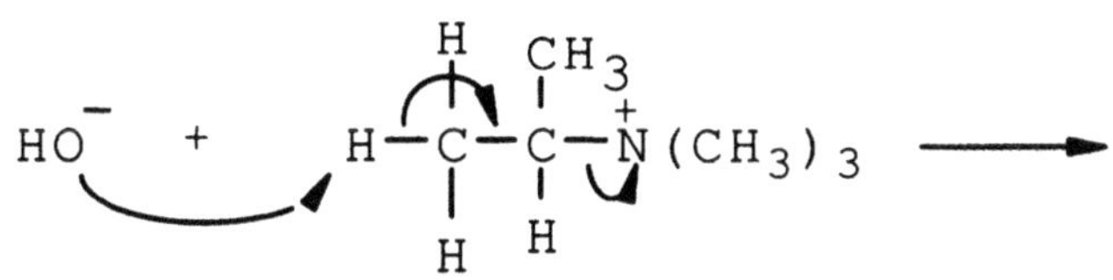

* *

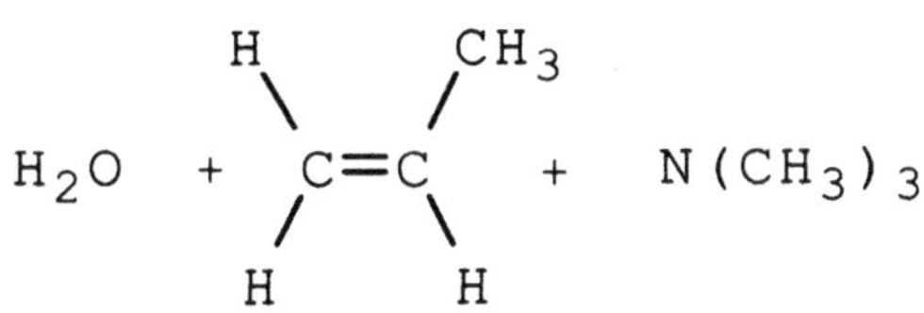

23

This is just another E2 reaction, like that in frames 11 and 12.

? *What will be the product(s) in the following reactions (first step only)?*

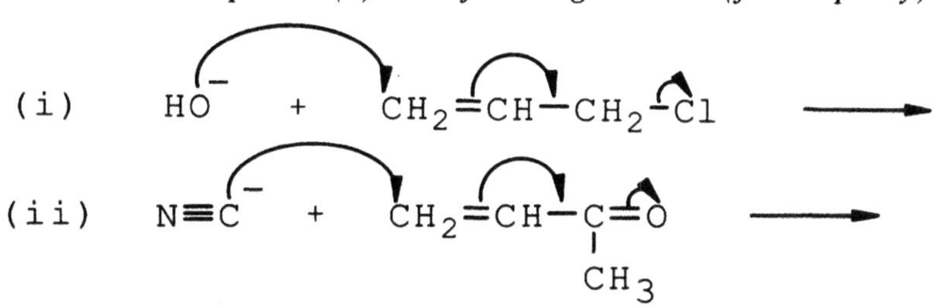

* *

(i) $HO{-}CH_2{-}CH{=}CH_2 \quad + \quad Cl^-$

(ii) $N{\equiv}C{-}CH_2{-}CH{=}C(CH_3){-}O^-$

24 The carbanion produced in (ii) is readily protonated as follows:

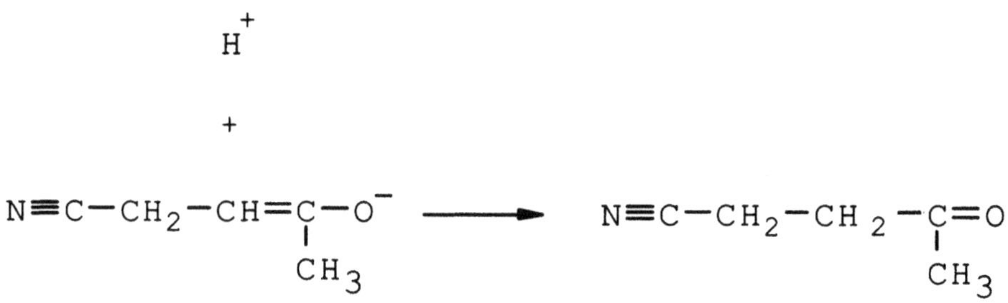

? *Rewrite this equation complete with the appropriate curly arrows.*

* *

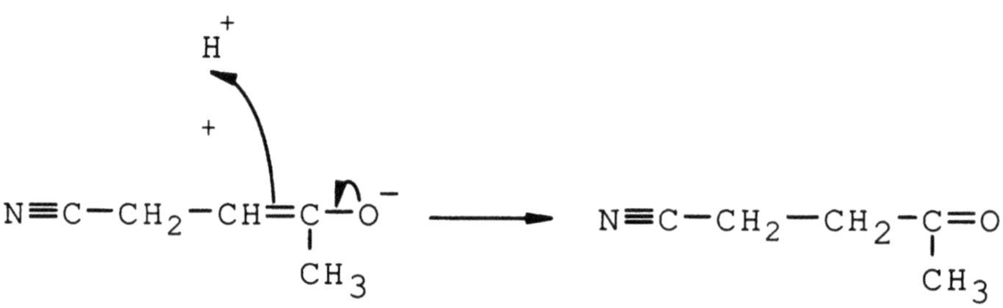

25

? *Now write in the curly arrows for the process given below. Note that we have switched to using the skeletal representation of an organic structure. This is a very convenient way of writing structures; there is a carbon at every junction between lines and at the end of every line. Bonds to hydrogen, and hydrogen atoms, are not shown.*

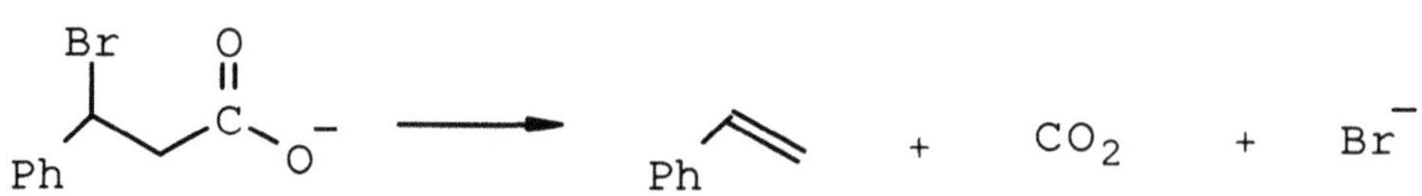

* *

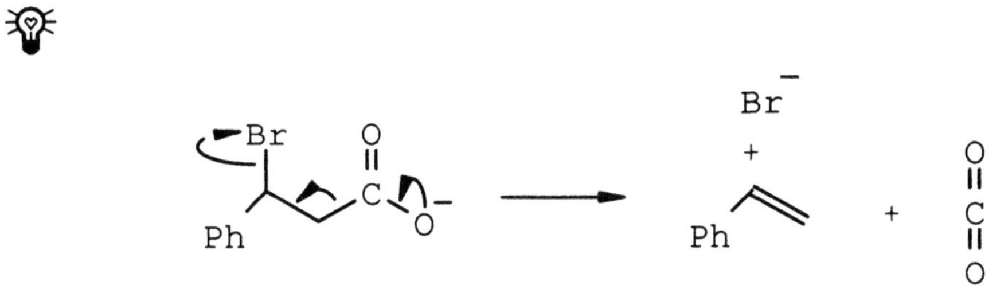

26 ? *Another problem in similar style – give the products for the following reaction:*

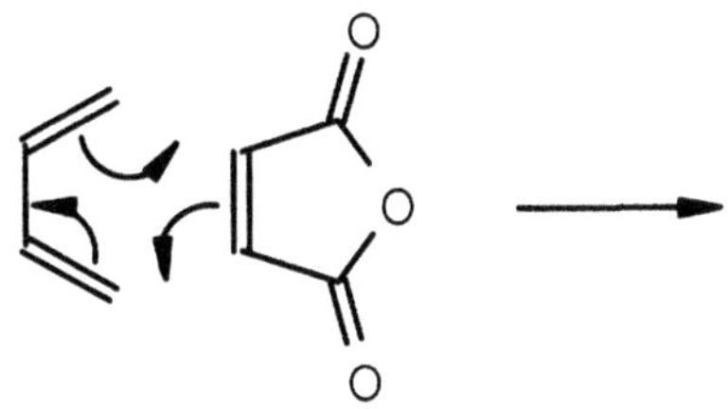

* *

A rather more difficult exercise is to come now, and it involves following through a complex multi-step reaction. The sequence below represents the formation of an oxime from a simple ketone and hydroxylamine. Occasionally some of the smaller products have been omitted; it is often convenient to do this and you should not be put off by it.

27

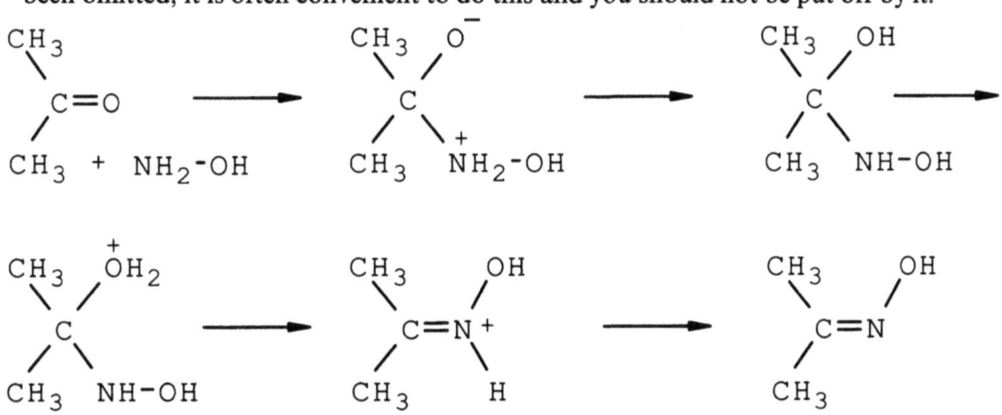

? *Rewrite the oxime formation sequence, complete with the appropriate curly arrows. It may be necessary to make slight alterations to the layouts of some of the structures to ensure that they display properly the features that you need.*

* *

28

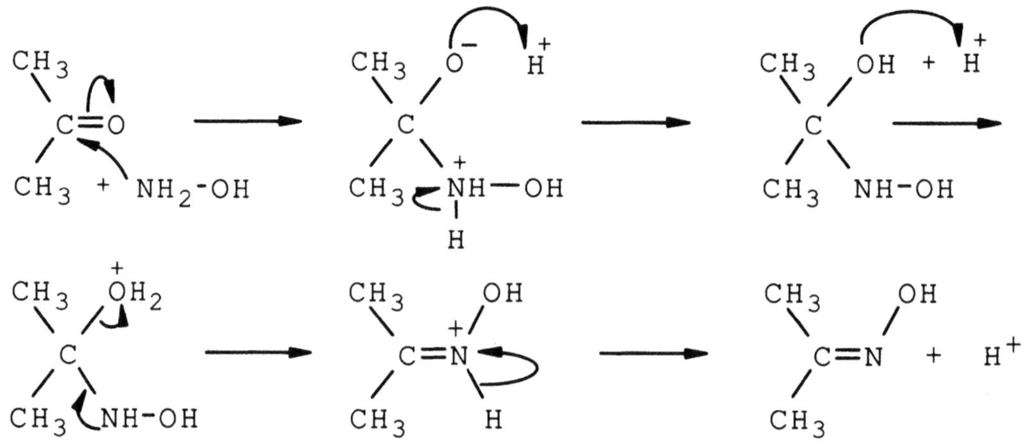

Note how one of the reacting N–H bonds had to be emphasized – or made more explicit – in order to put in the curly arrows, and how we had to bring up protons at appropriate points. They would not really be naked protons of course, but it is simpler to write them that way and to remember that they would, in reality, be part of an acid such as H_3O^+.

? *Try another, similar exercise, writing the curly arrows into the sequence:*

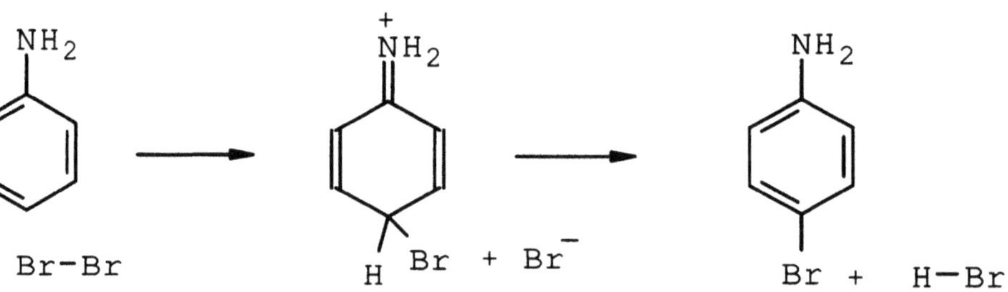

* *

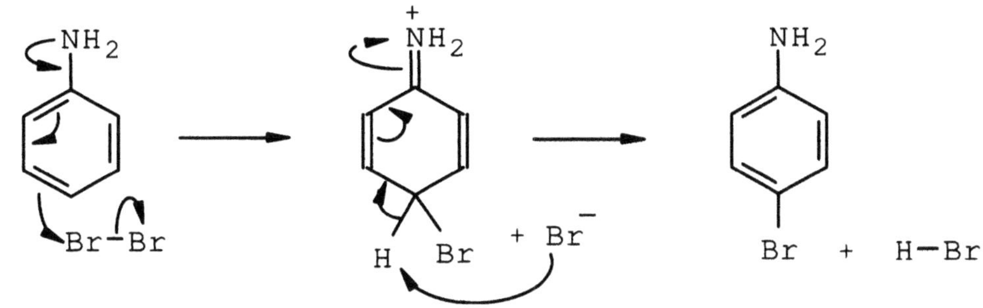

29 ? *Here is one final exercise of this sort. Supply the curly arrows for:*

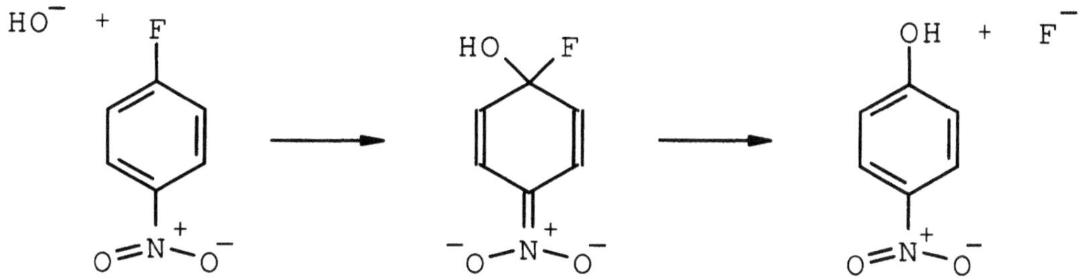

* *

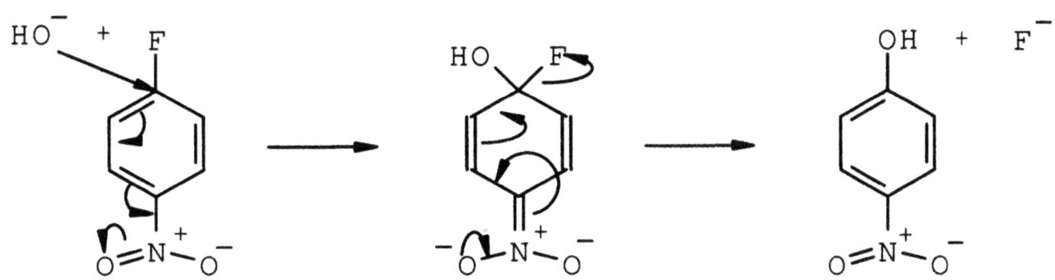

30

Finally, a couple of exercises which provide a little more to get your teeth into

? *Give the products for the following reaction:*

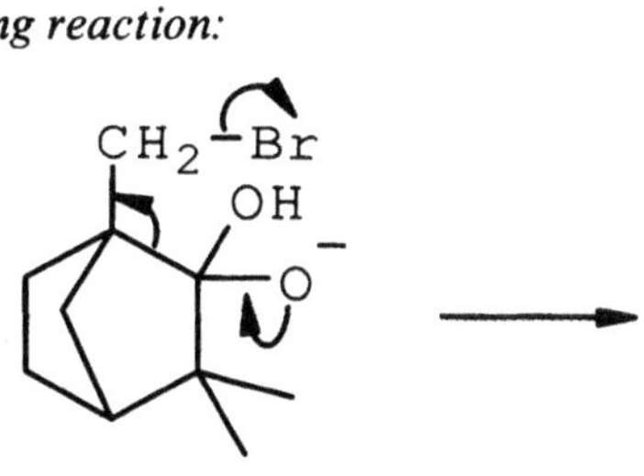

* *

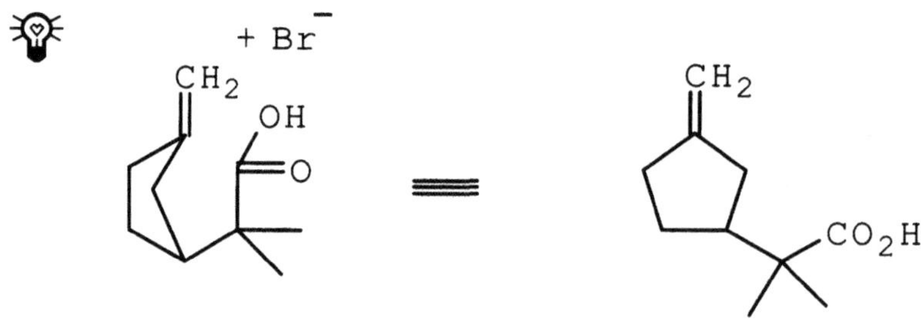

31

? *Give the curly arrows for:*

* *

32

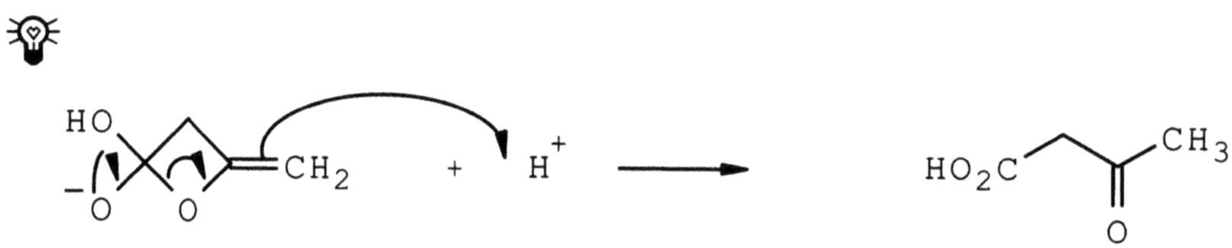

This is the end of this programme on curly arrows and organic reaction mechanisms, and you should by now be able to handle these ideas with some confidence. Mechanisms are likely to be written in this way throughout the whole of your course, and your skills in writing them should grow and develop. As this happens, you should also be able to develop the facility for making sensible predictions about the likely ways in which various possible species might react with each other.

* *

Programme 5

Dynamic stereochemistry – Part I

Most courses in organic chemistry introduce stereochemistry in terms of static molecules, defining terms such as configuration and conformation, and defining various sorts of stereoisomer. Analysis of particular compounds, in terms of how many and what sort of stereoisomers may exist for each one, is also covered. Whilst conformations are defined in a more or less static way, free rotation about single bonds (dynamic) in simple open chain molecules, and the consequent continuous interconversion of conformations, must have been understood.

When a particular stereoisomer undergoes chemical reaction to give a product which can exist in several stereoisomeric forms, the reaction may lead largely to one stereoisomer, in which case the reaction is said to be **stereoselective**. It may give exclusively one stereoisomer, in which case it would be of high (100%) stereoselectivity. If it gives comparable amounts of each stereoisomer, then it is said to show no (or very low) stereoselectivity.

If the reaction is reversible with respect to all possible product stereoisomers, then an equilibrium will be established amongst the product stereoisomers which will reflect their relative stabilities. The reaction itself will exert no control over the relative quantities of product stereoisomers, and is said to be under **thermodynamic control**. If the reaction is irreversible, however, then product stability will be of no direct importance since no equilibrium will be established. The relative amounts of product stereoisomers will be determined by the nature (mechanism) of the reaction itself, with the one which forms fastest predominating, or forming exclusively. The reaction is then said to be under **kinetic control**, and the stereochemistry of the product tells us something about the mechanism. Kinetic control is very common in organic reactions, and it is this sort of reaction which will be of most interest to us in this programme.

An example of a stereoselective reaction, under kinetic control, is the addition of bromine to cyclohexene. The product is (*E*)-1, 2-dibromocyclohexane, with the bromines *trans* to one another; no (*Z*)-isomer is formed with the bromines *cis*. The reaction mechanism determines the stereochemical outcome by imposing the rule, or constraint, that the two bromine atoms must approach the alkene molecule each from a different side of the double bond. This is known as the stereochemistry of the reaction; it is said to be a **trans addition**.

In this programme we shall be looking at the changes of stereochemistry which occur during the course of kinetically controlled chemical reactions. Our principal concern will be to learn to manipulate stereochemical diagrams on paper with a view to relating the configuration of products, the configuration of starting materials and the stereochemistry of the reaction. The essential prerequisites are a sound knowledge of static stereochemistry and the ability to draw good, clear, three-dimensional diagrams. We shall also reinforce some of the static stereochemistry which you have already learnt.

Bimolecular nucleophilic substitution–the S_N2 Reaction

The first reaction which we shall consider is the relatively simple S_N2 reaction, in which the configuration of the starting material is inverted at the carbon attached to the leaving group. You have come across this reaction before (Programme 4, frames 9 and 10).

The hydrolysis of 2-chloro-octane (leaving group attached to a chiral carbon atom) by aqueous sodium hydroxide (i.e. hydroxide ion), when second order kinetics are followed, is an example of this type of reaction:

$C_6H_{13}CH(Cl)CH_3 + OH^- \rightarrow C_6H_{13}CH(OH)CH_3 + Cl^-$

1 If a single, pure enantiomer of the starting material (or substrate) is used, then a single enantiomer of the product is obtained; the reaction is stereoselective and proceeds with inversion of configuration at carbon to give a product of configuration opposite to that of the substrate.

? *Draw a diagrammatic scheme to represent the stereochemistry of the reaction, starting from the given enantiomer, and laid out, from left to right, as:*

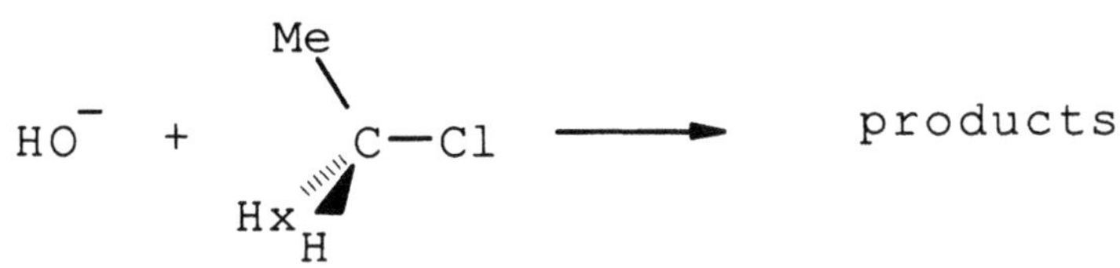

(Me = methyl, CH_3; Hx = Hexyl, C_6H_{13})

* *

Your scheme should look like:

2 Note how the incoming nucleophile approaches, and the leaving group departs from, the reacting carbon atom along the same 'line' but on opposite sides; the developing new bond and the decaying old bond are collinear, and the nucleophile is said to attack the substrate molecule from the rear side.

It is also important to note that it is usual to draw the diagrams so that the line of the newly developing and the decaying bonds lies horizontally in the plane of the paper. This orientation is the most convenient one to draw, and it makes for the clearest possible diagram.

This is a one step reaction, and there are no intermediates involved in the pathway. It is useful, though, to consider the **transition state** for this single step in which the new and the old bonds are each partially formed. The transition state represents the nature of the reacting species at the top of the reaction profile, and the difference in energy between the ground state (reactants) and the transition state is the activation energy of the reaction, a major factor in determining the reaction rate. A consideration of the transition state of the rate determining step of a reaction, compared with the ground state, can help us to understand how the rate of the reaction is influenced by a variety of factors.

? *Draw a diagram of the transition state for this reaction.*

* *

Your diagram should look like:

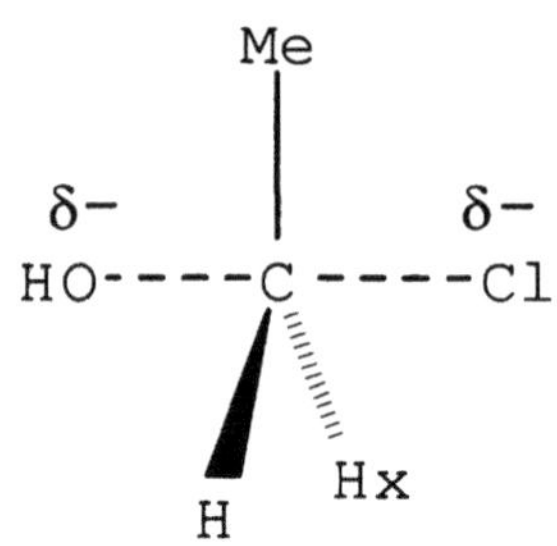

Note that the central carbon and the three 'non-reacting' groups (Hx, Me and H) are coplanar, and the charge originally associated with the attacking nucleophile, OH^-, is equally divided between the leaving and entering groups.

3 Strictly speaking, this is only true if the leaving group and the entering group are identical; only in this situation would the transition state be perfectly symmetrical about a plane orthogonal (perpendicular) to the developing/decaying bonds and containing the reacting carbon atom. However, it is a good approximation for our purposes.

We might also add at this point that the two partial negative charges may sum to one whole negative charge, or to more than one. In the latter case the carbon will bear a small positive charge; the net charge on the transition state must be a whole negative charge since it must equal the net charge on the reactants.

The carbon can never bear a negative charge, though, with the sum of the two partial negative charges on the entering and leaving groups less than one, as that would imply more than eight electrons in the valence shell of carbon. It can 'hold' fewer than eight, but never more.

? *Draw a scheme incorporating reactants, transition state and products, to show the complete reaction and laid out as:*

reactants ⟶ *transition state* ⟶ *products*

Draw arrows over or under the 'non-reacting' groups attached to the reacting carbon atom in the transition state to show their direction of movement relative to that carbon.

* *

Your scheme should represent the line of developing and breaking bonds horizontally, and should look like:

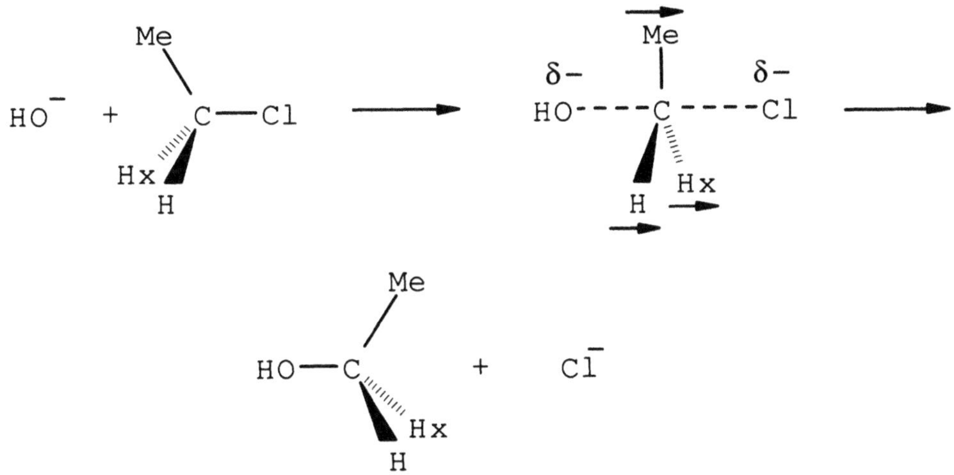

4

Note that the arrows associated with Me, Hx and H show movement from left to right in the diagram – the 'direction' of reaction – and illustrate the inversion of the chiral centre, which is rather like the blowing inside out of an umbrella in a strong wind.

? *Now draw a diagrammatic scheme to represent the stereochemistry of a different nucleophilic substitution, reduction of 2-chloro-2-deuterio-2-phenylethane. In this reaction the hydride ion (the nucleophile) displaces the chloride ion (the leaving group). Set the reaction out as in frame 2:*

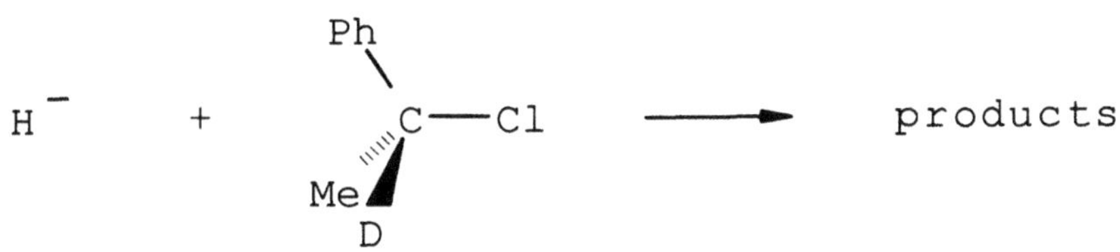

(Ph = phenyl, C_6H_5)

* *

5

Your scheme should look like:

? *Work out the absolute configuration (**R** or **S**) of both the reactant halide and the product alcohol for the reaction in frames 3/4.*

* *

The halide is S *and the alcohol* R.

If you are in doubt about this, check that you have allocated the priorities of the groups attached to the chiral centre correctly, and that you have 'looked' at the molecules from the chiral centre along the bond to the group of lowest priority (hydrogen in both molecules).

6 The priorities should have been:

for the halide: Cl, 1; Hx, 2; Me, 3; H, 4

for the alcohol: OH, 1; Hx, 2; Me, 3; H, 4.

? *Now work out the absolute configurations of the reactant halide and the product deuterated hydrocarbon for the reaction in frames 4/5.*

Both the halide and the hydrocarbon are of the R *configuration.*

The priorities should have been:

for the halide: Cl, 1; Ph, 2; Me, 3; D, 4

for the hydrocarbon: Ph, 1; Me, 2; D, 3; H, 4.

Compare the two reactions (frames 3/4 and 4/5) in terms of the absolute configurations
7 of reactant and product. In the halide hydrolysis (top of frame 4) we have a change from *S* to *R*; in the case of the halide reduction (top of frame 5) we have no change – both reactant and product are *R*.

? *For the second reaction, is there a conflict between the statement that the absolute configurations of the reactant and the product are the same, but the reaction has proceeded with inversion of configuration?*

No.

Remember that the symbols *R* and *S* are decided by reference to a scale of priorities which is arbitrary. Any convenient criterion of allocating priority could have been chosen; it did not have to be atomic number.

In the frame 3/4 reaction, the entering and the leaving groups both had the same priority (1), and inversion of configuration during the reaction shows up nicely as a change from *S* to *R*. In the frame 4/5 reaction, however, the entering group (H) has a lower priority
8 (4) than the leaving group (1), and this change in priority means that the specifications of absolute configuration (*R* or *S*) cannot be directly compared. The product (*R*) and the reactant (*R*) are of opposite configuration in the spatial sense (the 'real' sense).

Had we been given the absolute configurations (as *R* or *S*) of the reactants and products of the frame 4/5 reaction, it would have been necessary (in practice it is always wise) to work out the three-dimensional disposition of the groups attached to the chiral centres, in reactant and in product, in order to decide whether the reaction has proceeded with an inversion of configuration, or with retention. Had both reactants and products been of the same three-dimensional configuration, one would have been *R* and the other *S*.

Unimolecular nucleophilic substitution – the S_N1 reaction

Consider now the stereochemical course of an S_N1 reaction. The 'ideal' S_N1 reaction proceeds via a carbocation intermediate which is trigonal and perfectly symmetrical about the flat plane which characterizes it.

We could represent such a cation, derived from a chiral substrate CabcL (L is the leaving group), as follows. Note that we are looking at it 'edge-on'.

9

? *Draw a diagram representing the result of a nucleophile attaching to this cation after attacking it FROM THE LEFT SIDE. Assume that the nucleophile is Nu^-.*

* *

Your product should be:

10

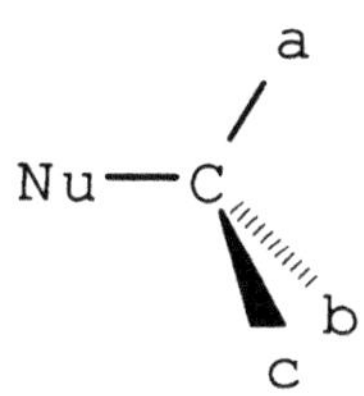

? *Now draw a diagram representing the product obtained from attack on the cation FROM THE RIGHT SIDE.*

* *

Your product should be:

11

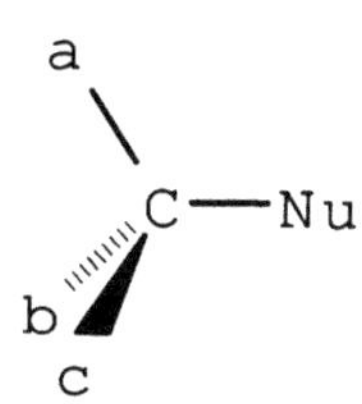

? *What is the relationship between these two products which you have just drawn (frames 10 and 11)?*

* *

One is the enantiomer of the other.

12 ? *Bearing in mind what was said (frame 9) about the symmetry of the cation, what proportions of each of these two species would you expect to be formed by the attack of Nu⁻ on a quantity of the cation?*

* *

13 *They would form in equal quantities, since there must be an equal probability of the attack occurring at either side of the symmetrical cation.*

? *How, then, would the product of this reaction be described?*

* *

14 *It would be described as a racemic mixture.*

The product, then, of an ideal S_N1 reaction at a chiral centre will be a racemic mixture. The reaction is said to proceed with **racemization**, and this is a characteristic of an ideal S_N1 reaction.

An example would be the following methanolysis reaction:

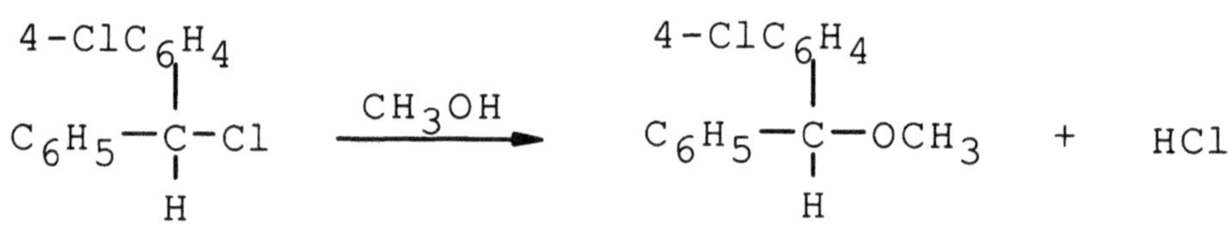

(laevorotatory) (optically inactive) (racemic)

Not all S_N1 reactions are 'ideal', however, and some proceed with less than 100% racemization. They are characterized by a certain amount of inversion.

An example is the S_N1 methanolysis of 1-bromo-1-phenylethane:

$$C_6H_5CH(Br)CH_3 + CH_3OH \longrightarrow C_6H_5CH(OH)CH_3 + HBr$$

R-configuration

63% *S*-configuration (inverted)
37% *R*-configuration
(configuration retained)

Note that the product can be regarded as being made up of 74% of the racemic mixture (the 37% of product which is *R*, together with 37 of the 63% which is *S*) and 26% (63–37 = 26) of the *S* enantiomer. The reaction has thus proceeded with 26% inversion and 74% racemization.

? *What conclusion can be reached from this result, given that the reaction is certainly S_N1?*

* *

The conclusion has to be that the intermediate which is reacting with the nucleophile is not completely symmetrical, so that attack at one side is preferred, to a degree, over attack at the other side.

In the 'ideal' S_N1 reaction, which we considered above, the carbocation was said to be completely symmetrical about its own plane. This would be true if it were completely free. It would be solvated, of course, but in a way which reflected its own symmetrical nature. It may, however, have a very short lifetime; the nucleophile may attack it before the leaving group has completely disappeared from its proximity. Even though not bonded to the cationic centre, the leaving group would then partly cover one side of the cation and predispose the nucleophile to attack from the other side. This, of course, would lead to rather more of the inverted product than of the product with retained configuration.

15 This is how the results observed with 1-bromo-1-phenylethane may be explained.

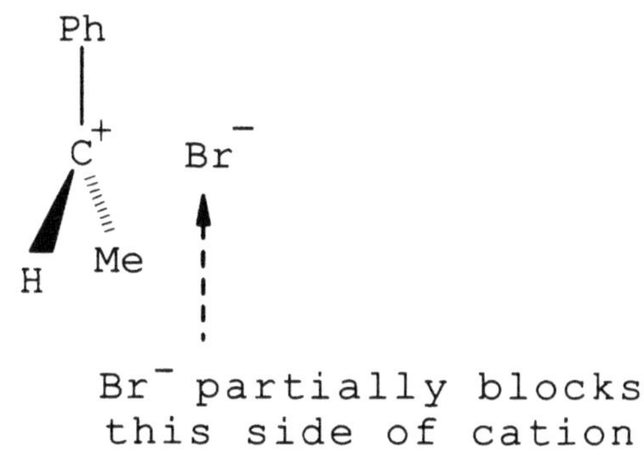

Nucleophilic substitution in a different context: opening of epoxide rings

Epoxide rings, by virtue of their high ring strain, are very readily opened when attacked by a nucleophile. The nucleophile replaces the ring oxygen in its attachment to one of the epoxide carbon atoms:

16

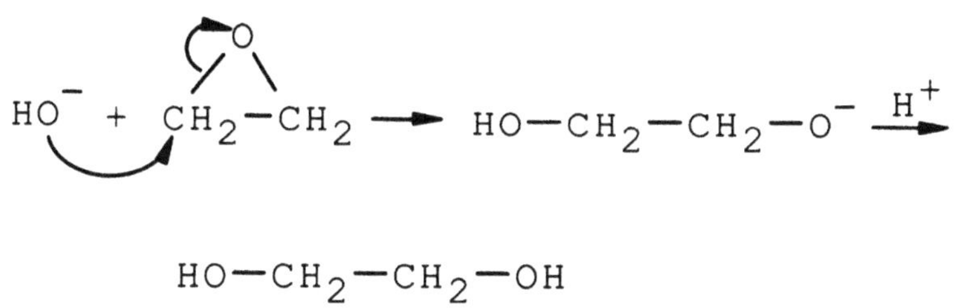

? *To what mechanistic class does this reaction belong?*

* *

This is an S_N2 reaction.

The nucleophile is the hydroxide ion, and the leaving group is the $-CH_2-O^-$ group. Reactions of this type may be either S_N1 or S_N2. In this case the reaction is S_N2.

S_N2 reactions of this type will be stereoselective, of course, whereas S_N1 reactions will not, so we are principally interested in the S_N2 reactions here.

17 ? *By means of suitable diagrams, trace the stereochemical course of the S_N2 ring opening, by hydroxide ion, of the epoxide (Z)-2-butene oxide, and predict the stereochemistry of the product. HINT: keep the reacting atoms/groups in the plane of the paper for simplicity and clarity.*

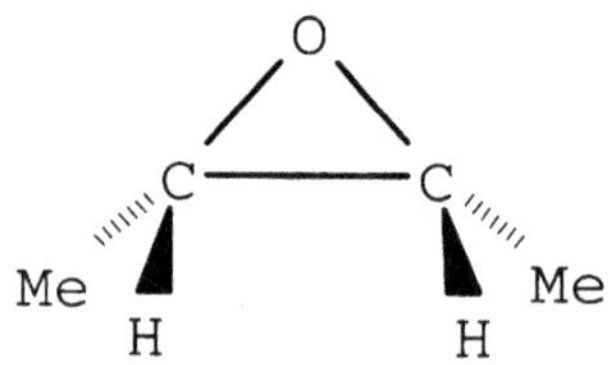

(Z) -2-butene oxide

* *

Starting with attack at the left-hand carbon, your scheme should look like:

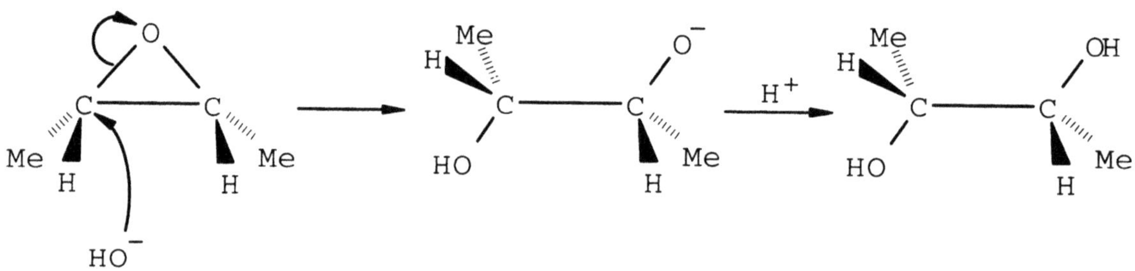

18 *But note that attack at the right-hand carbon is equally likely:*

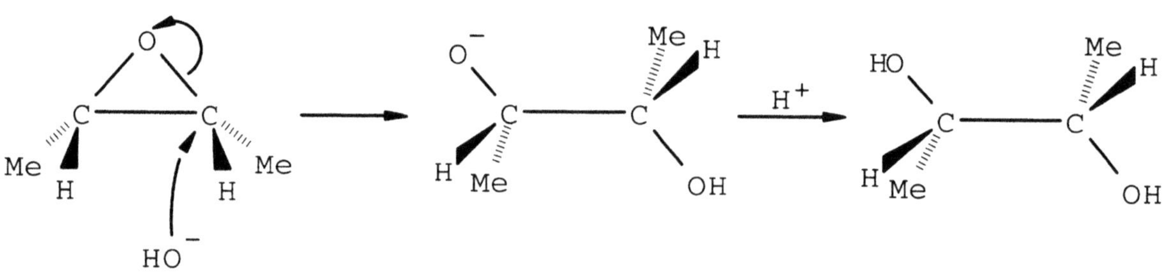

The two products are non-superimposable mirror images of one another (imagine a 'horizontal mirror' between the two schemes). They will be produced in equal amounts since attack at either of the epoxide carbons is equally likely. In other words it is *racemic* 2, 3-dihydroxybutane which is produced.

A different product would be obtained, of course, if the starting epoxide had been (*E*)-2-butene oxide.

? *Use your (developing) chemical intuition (i.e. without using diagrams) to predict the product which would be obtained from the (*E*)-epoxide.*

* *

19

The expected product would be the meso-*diastereoisomer of the diol.*

Each of the two chiral carbons in the diol has the same set of four different groups attached to it, so that the diol may exist in three stereoisomeric forms, a (+)-form, a (−)-form (racemic when mixed in equal proportions) and a *meso*-form. The racemate would be diastereoisomeric with the *meso*-form, and it would be reasonable to expect the diastereoisomeric product to form if one changes the substrate from the (*Z*)-form to the (*E*).

? *Trace the stereochemical course of the ring opening of the (*E*)-epoxide, and show that the product is, indeed, the* meso-*form.*

* *

20

Your answer should be as follows:

Attack at the left-hand carbon:

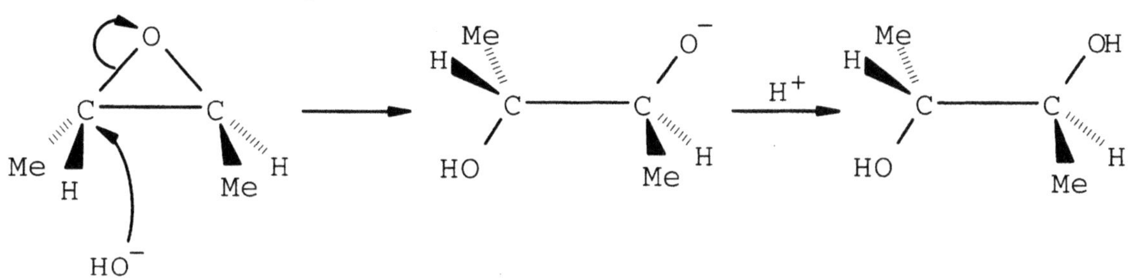

Attack at the right-hand carbon (equally likely):

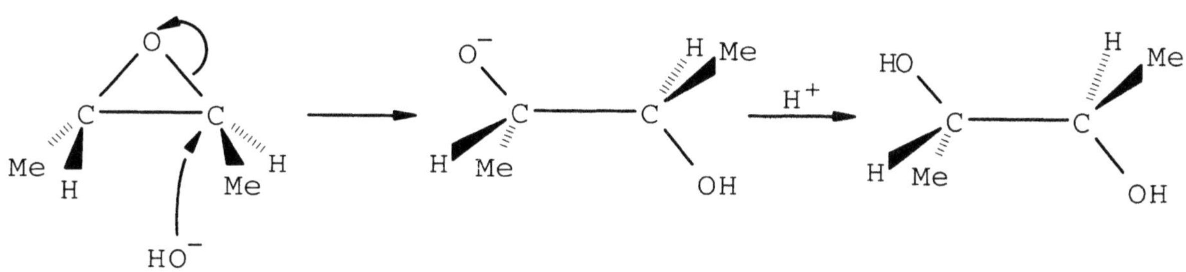

Note that the product is the same, the *meso*-form, whether attack takes place at the right or the left-hand carbon.

The synthesis of an epoxide is also achieved by a stereoselective reaction, the oxidation (epoxidation) of an alkene with a peroxyacid:

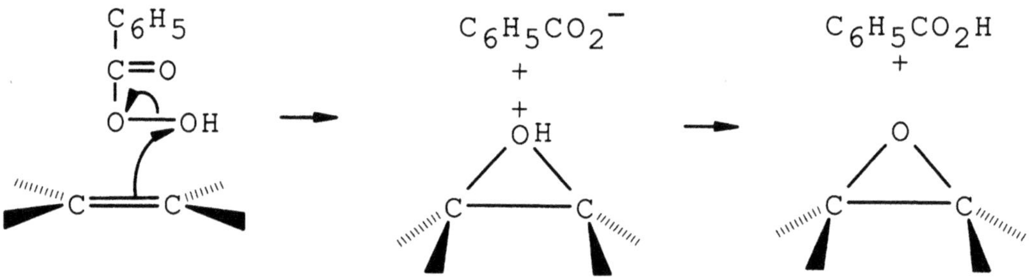

? *What would be the stereochemistry of the epoxide (2-butene oxide) obtained by epoxidation of (E)-2-butene with peroxybenzoic acid?*

* *

21 *The epoxide would be (E)-2-butene oxide.*

Since the epoxide oxygen (frame 20) has to approach the alkene from one side, the stereochemistry ((*Z*) or (*E*)) of the alkene will be preserved in the product.

Neighbouring group participation

A phenomenon which is in some ways related to the opening of epoxide rings is **neighbouring group participation**. This is a concept used to explain the fact that molecules such as 2-chloroethyl ethyl sulphide will undergo nucleophilic substitution several thousand times faster than the corresponding chloro-alkane (1-chloropentane).

The idea is that the neighbouring sulphur atom, because it is so close to the reacting carbon and because it is such a good nucleophile, readily displaces the chloride ion to form a strained three-membered ring, and that the attacking nucleophile can then rapidly open the ring.

Thus, for hydrolysis:

22

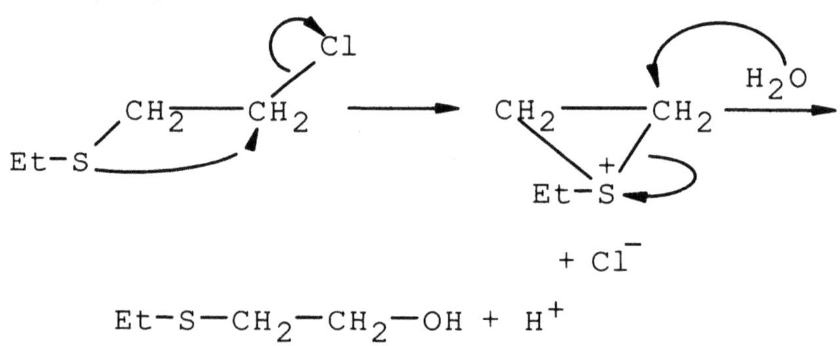

? *Given this mechanism, predict the stereochemistry of the hydrolysis of the chiral substrate below:*

$$\text{Et-S-CH}_2\text{-CH(Me)-Cl} \xrightarrow{H_2O}$$

$$\text{Et-S-CH}_2\text{-CH(Me)-OH} \quad + \; H^+ \; + \; Cl^-$$

* *

23 *Since the reaction proceeds with two S_N2 reactions at the reacting carbon, each of which must involve inversion, the overall result will be retention of configuration.*

One inversion followed by another will return the configuration of the reacting carbon to that in the substrate.

? *Take the* S *enantiomer of the substrate as an example, and follow the reaction through, with standard stereochemical diagrams, and confirm this conclusion.*

* *

24 *Your answer should be:*

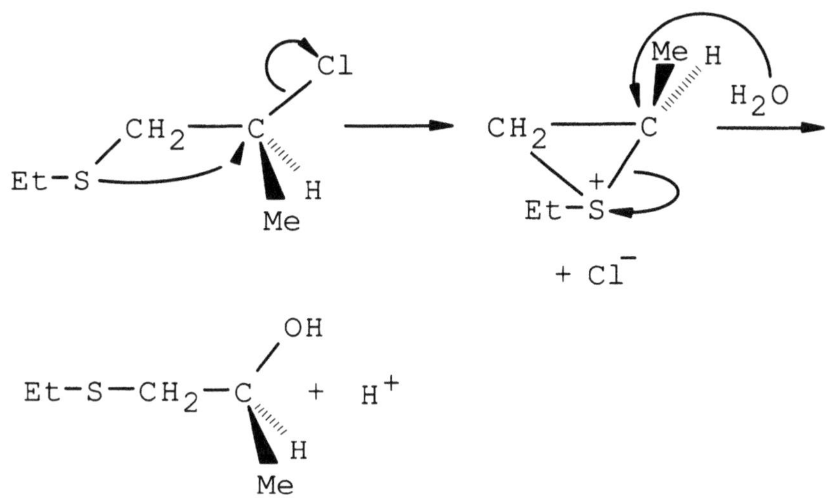

Electrophilic addition to alkenes

The stereoselective addition of bromine (in an inert solvent) to an alkene is another reaction related to epoxide ring opening. The *anti*-addition proceeds *via* the transfer of a bromonium ion (Br^+, an electrophile) from a bromine molecule to the alkene (like the epoxidation of alkenes, where HO^+ is transferred from a peroxyacid to the alkene), and the resultant cyclic bromonium ion is opened by the remaining bromide ion, a nucleophile. It is an example of electrophilic addition.

25

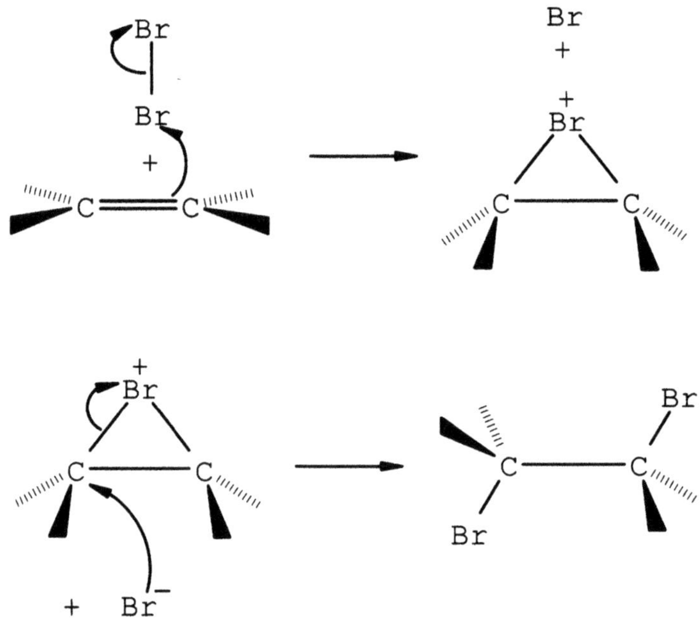

? *Predict the stereochemistry of the product obtained by the electrophilic addition of bromine to (Z)-butenedioic (maleic) acid.*

* *

The product will be racemic 2,3-dibromobutanedioic acid (2,3-dibromosuccinic acid).

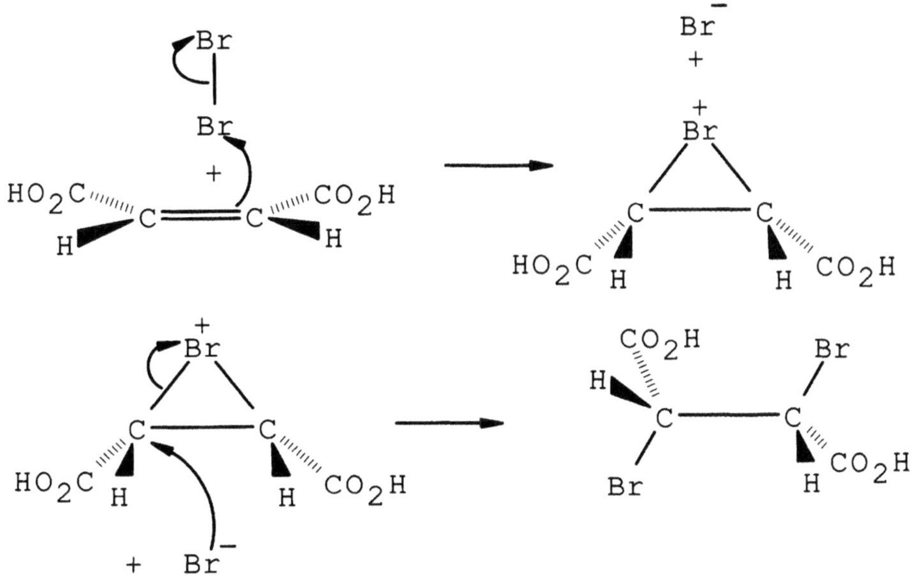

26

Attack at the right-hand carbon of the cyclic bromonium ion would lead to the enantiomer of the product shown (frame 18), in an equal quantity.

In these cases the product is shown 'frozen' in the conformation in which it was initially formed. Rotation about the central C–C bond would soon destroy the *anti*-relationship between the bromines, but would not, of course, alter the **configurations** of the chiral centres. They would remain unchanged and would demonstrate the *anti*-nature of the addition.

There is no need to write down the cyclic bromonium ion intermediate, of course, merely to work out the stereochemistry of the product. It would be sufficient to know that the addition is *anti*.

? *Without writing down the intermediate cyclic bromonium ion in your scheme, predict the stereochemistry of the product obtained from the* anti-*addition of bromine to cyclopentene.*

* *

*The product would be (*E*)-1,2-dibromocyclopentane.*

27

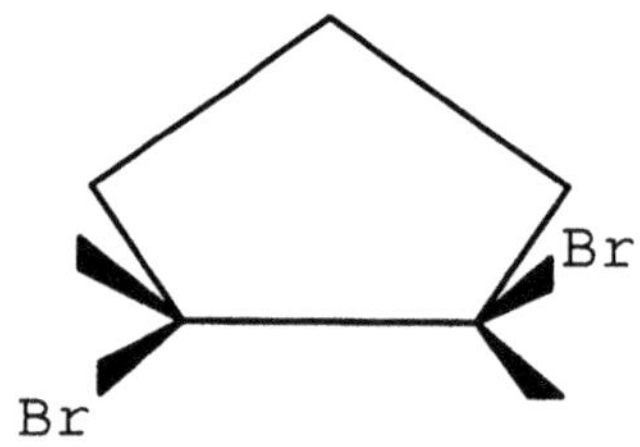

Note that the *anti*-relationship between the bromine atoms is preserved in the product.

? *Why should this be so?*

* *

Because the ring structure prevents rotation about the reacting C–C bond after it has become a single bond, the conformation initially produced about that bond must be preserved as a reflection of the anti-*nature of the addition reaction.*

28

Having noted the similarity between the epoxidation of alkenes followed by nucleophilic ring opening on the one hand, and both neighbouring group participation and electrophilic addition on the other, the isolation of stable epoxide products from the epoxidation reaction can be seen to provide supporting evidence for the involvement of unstable (non-isolated) three-membered ring intermediates in the other two. The stereochemical similarities between all three reactions also support the idea that cyclic intermediates are involved in all three.

* *

This is the end of Programme 5, Dynamic Stereochemistry Part I. When you are confident that you have mastered the knowledge and skills within this programme, proceed on to Programme 6, Dynamic Stereochemistry Part II. If there are any difficulties which you cannot resolve from the Programme itself, then seek advice from a tutor.

Programme 6

Dynamic stereochemistry – Part II

Bimolecular elimination – the E2 reaction

We now move on to consider a different type of reaction, 1,2-elimination, in particular 1,2-eliminations which proceed by the E2 mechanism.

Take the base-catalysed E2 elimination of HBr from 2-bromo-3-methylpentane to give 3-methyl-2-pentene.

1

$$C_2H_5CH(CH_3)CH(Br)CH_3 \rightarrow C_2H_5C(CH_3)\text{=}CHCH_3 + \text{'HBr'}$$

There are two different chiral centres in the substrate and, in this reaction, both are involved as reacting centres.

? *How many stereoisomers are there for the substrate in this reaction?*

* *

There are four possible stereoisomers for the reactant bromide. There are two different chiral centres, hence 2^2 stereoisomers.

There is one pair of enantiomers with the configurations [2(*R*),3(*R*)] and [2(*S*),3(*S*)], and a diastereoisomeric pair with configurations [2(*S*),3(*R*)] and [2(*R*),3(*S*)].

2 ? *Copy out the partial structure (the 'blank'), given below, of the [2(*R*),3(*R*)] isomer, and then complete the diagram by filling in the missing groups a, b, c, and d.*

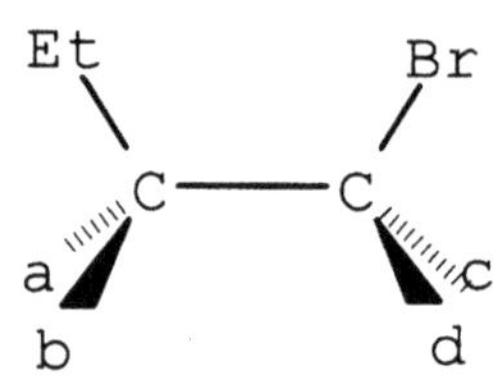

* *

Your completed diagram should be as follows:

3

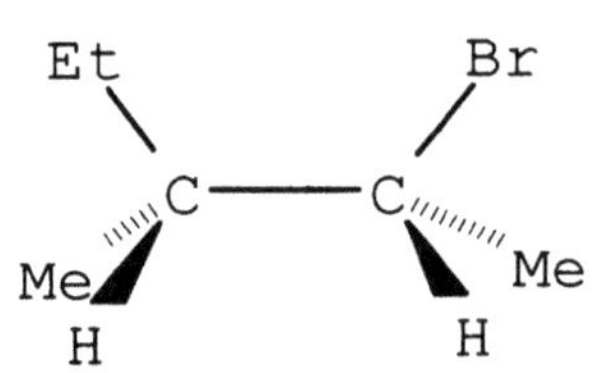

? *Now draw, by specifying (differently!) the groups a, b, c and d on the 'blank' given in frame 2, the stereoisomer of the substrate compound with configuration [2(*R*),3(*S*)].*

* *

Your completed diagram should be as follows:

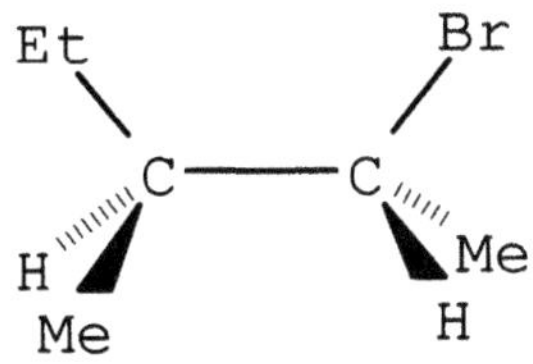

4 Notice how, in the [2(*R*),3(*R*)] stereoisomer (top of frame 3), the methyl groups eclipse each other, as do the hydrogen atoms, in the particular conformation drawn. The same would, of course, be true of its enantiomer [2(*S*),3(*S*)], and these two stereoisomers are the enantiomers of what is known as the *erythro*-form.

In the diagram of the [2(*R*),3(*S*)] stereoisomer, the methyls are not eclipsing each other, and neither are the hydrogens. If you imagine rotation about the central C–C bond until the hydrogens eclipse each other, then the methyls would not eclipse each other. If you imagine rotation about the C–C bond until the methyls eclipse each other, then the hydrogens would not. No conformation is possible in which the identical groups all eclipse one another. This stereoisomer and its enantiomer (of which the same is true) are known as the *threo*-form. The *erythro*- and *threo*-forms are diastereoisomers, and they have different chemical properties (albeit not *very* different).

? *How many different stereoisomers are possible for the product alkene, and what type of stereoisomerism do they show?*

* *

5 *There are two stereoisomers of the product alkene, 3-methyl-2-pentene, and they are geometrical isomers. (Two diastereoisomers would also be a correct response to this question.)*

? *Draw out each of these two stereoisomers and label them with their absolute configurations.*

* *

6 *The two stereoisomers are:*

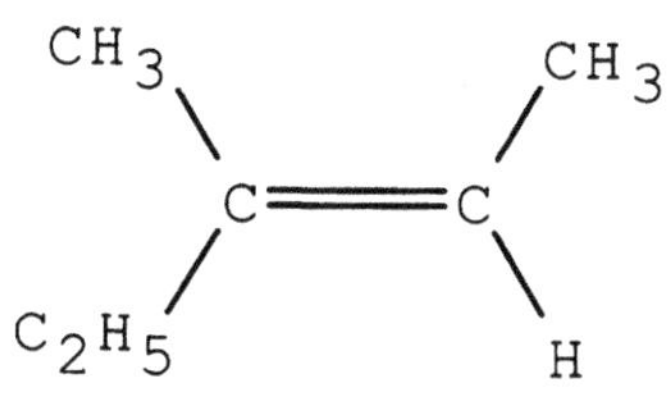

(E)-3-methyl-2-pentene

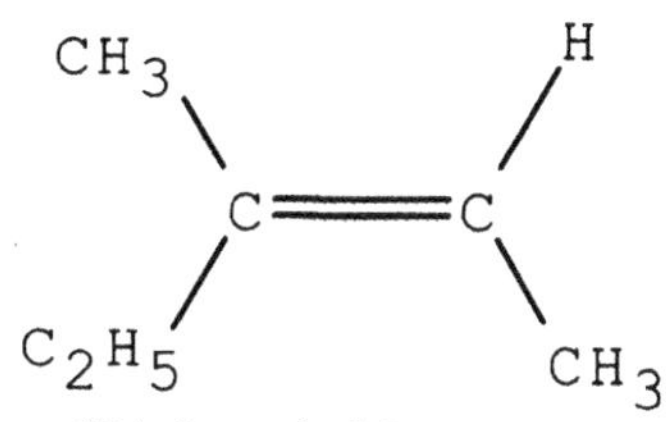

(Z)-3-methyl-2-pentene

In practice it is found that E2 elimination of HBr from the *erythro*-reactant (either enantiomer) gives only the (*E*)-alkene, and E2 elimination of HBr from the *threo*-reactant (either enantiomer) gives only the (*Z*)-alkene.

? *What is the term used to describe a reaction with characteristics like these? (Programme 5, frame 1).*

* *

The reaction is **stereoselective**, *since one stereoisomer of reactant gives only one stereoisomer of the product.*

Take one of the enantiomers of the *erythro*-form [2(*R*),3(*R*)], and consider how it can give rise, stereoselectively, to the (*E*)-alkene in which the two methyl groups are *cis*- to one another, as are the hydrogen and the ethyl group.

In the transition state of this one-step reaction, a double bond is developing between C2 and C3, so that rotation about that bond is already restricted, and the restriction is preserved (in fact enhanced) in the product alkene. The conformation of the transition state must therefore reflect the disposition of the substituent groups in the product, and this disposition must also be reflected in the reacting conformation of the substrate.

The transition state and products (the base catalyst is HO^-) may be represented as:

7

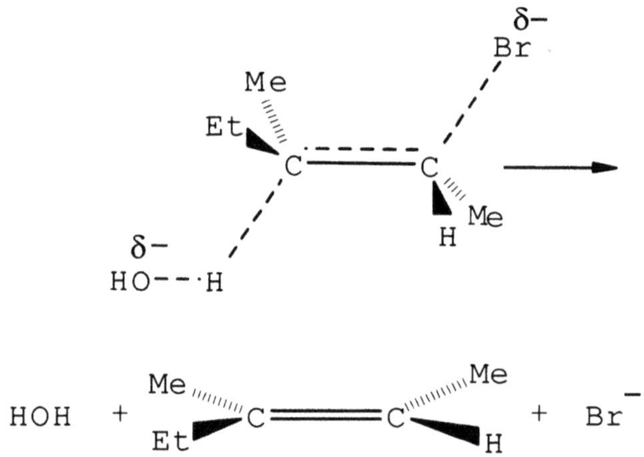

Notice how the product alkene has been drawn as an 'edge on' view, and that the reacting groups (HO^-, H, C–C, Br) in the transition state leading to it, all lie in one plane perpendicular to the plane of the product. Notice, too, how, for convenience and for maximum clarity, we have represented the plane containing the reacting groups (plane of reaction) in the transition state in the plane of the paper. This is always the clearest way to draw them. The substituents lie on either side of this 'plane of reaction' in a manner which reflects their stereochemical disposition in the product. Thus, the two methyl groups become *cis*-related in the product because they lie on the same side (far side in this diagram) of the 'reacting plane' in the transition state, and groups which lie on opposite sides of this plane, such as the ethyl group and the methyl group on C2, become *trans*-related in the product.

? *Draw a reaction scheme like the one above, but incorporating the hydroxide ion and the reacting conformation of the substrate to the left of the transition state. Remember that the reacting conformation of the substrate should reflect the spatial distribution of the substituent groups of the transition state, and, therefore, also of the product. Include appropriate curly arrows with the reactants on the left.*

* *

Your scheme should be as follows:

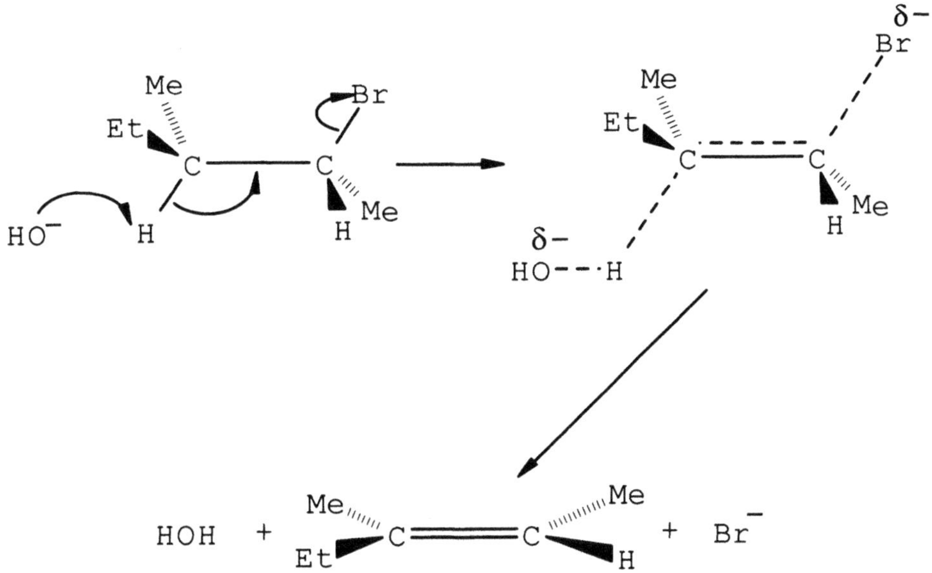

8 In both the reacting conformation of the substrate, and in the transition state, the departing hydrogen and bromine atoms lie in the plane of reaction, but on opposite sides of the C–C bond. They are said to be **anti-periplanar**. Except in certain rather special cases, E2 reactions follow this *anti*-periplanar rule.

Note how, had we started by considering the [2(*R*),3(*R*)] isomer of the substrate in the conformation represented in frame 3, it would have been necessary to rotate the 'left-hand' C atom about the C–C bond to get the molecule into this reacting conformation.

Had the departing groups lain on the same side of the C–C bond (still in the same plane) in the reacting conformation, they would have been described as **syn-periplanar**. There are some E2 reactions (frames 20 and 21) which do indeed adopt a *syn*-periplanar arrangement of reacting groups if an *anti*-periplanar arrangement is not possible. Coplanarity of the reacting groups is essential; an *anti*-periplanar arrangement is preferred (strongly) but a *syn*-periplanar arrangement is possible. The *syn*-periplanar arrangement, of course, represents a less stable, eclipsed conformation whereas the *anti*-periplanar arrangement represents a more stable staggered conformation.

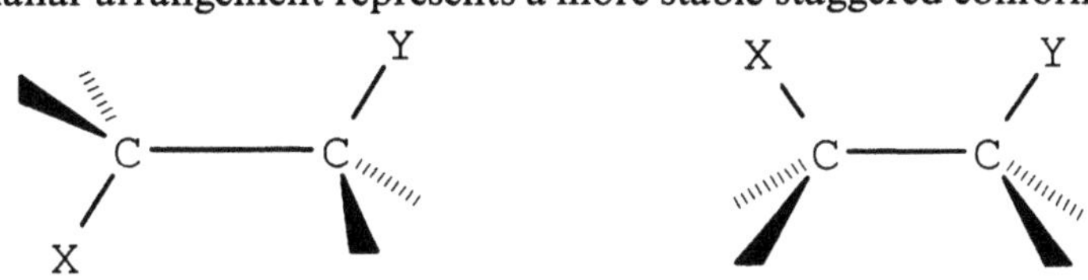

anti-periplanar *syn*-periplanar

X & Y represent
the departing
groups

It is not always the case that the reacting molecule which we may be asked to consider is presented in the proper conformation for reaction. It must therefore be subjected to rotation about the appropriate bond to get it into the proper reacting conformation before any predictions can be made about the stereochemistry of the products.

? *Starting from the chlorohydrocarbon given below, follow through the stereochemistry of the E2 elimination of hydrogen chloride to give 1,2-diphenylpropene, in such a way as to enable you to predict the stereochemistry of the product.*

8 *Do not include the transition state, since the prediction can be made directly from the reacting conformation of the reactants. If in doubt on this point, look again at frame 8 and recognise how similar the reacting conformation of the substrate is to the transition state.*

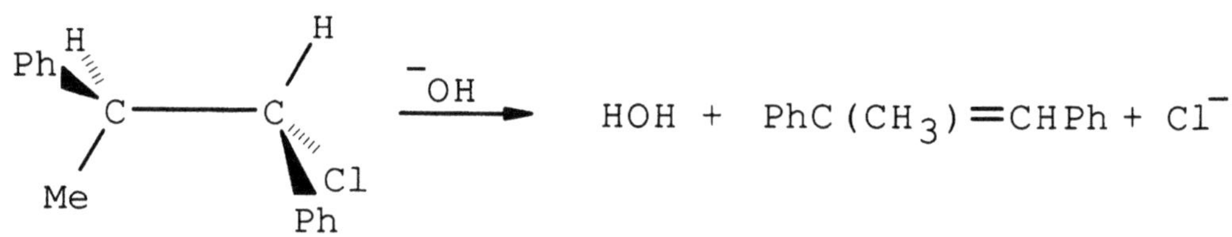

* *

Your diagrams should be as follows, leading to a prediction that the product would be (Z)-1,2-diphenylpropene:

9

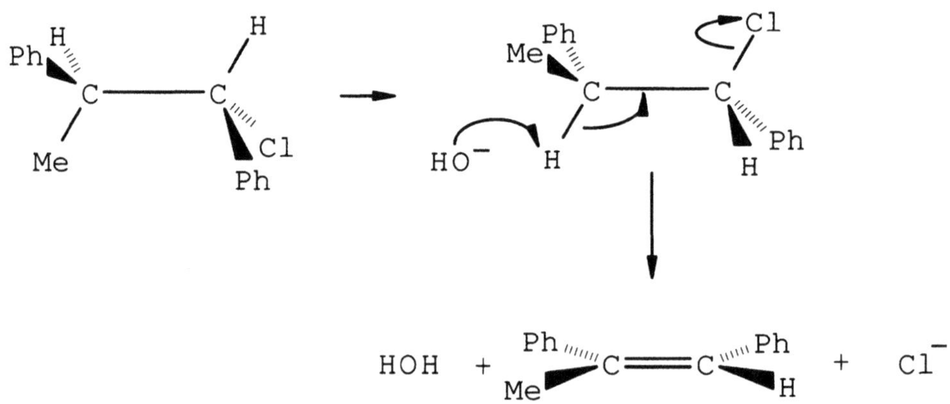

The first task in the solution to this problem is to recognize that:
(a) the initial conformation cannot lead to reaction since the departing groups and the two carbons destined to become the alkene double-bond carbons are not *anti*-periplanar, and
(b) neither of the bonds to the departing groups is in the plane of the paper.

It therefore follows that rotations about the C–C bond must be invoked until (a) the reacting groups are *anti*-periplanar and (b) the 'plane of reaction' corresponds to the plane of the paper. This is what is represented by the first step in the sequence above.

The second step in the sequence is the reaction itself, in which we can see that, in the product, the two phenyl groups must be *cis*-related, and the product must be the (*Z*)-alkene. Note that, in the reacting conformation as shown, the two phenyl groups are on the same (remote) side of the 'plane of reaction' (beneath the plane of the paper) and end up *cis*- to each other, while the hydrogen atom and the methyl group also end up *cis*- since they, too, are both on the same (this) side of the 'plane of reaction' (above the plane of the paper).

Occasionally, one may be asked to make a prediction about the outcome of an E2 elimination of a chiral substrate which has not been presented in a three-dimensional diagrammatic form at all. It may, for example, have been presented as a Fischer projection, or as a name in which the configurations of the chiral centres have been specified (as *R* or *S*).

In such a case it is useful to proceed through the following steps.

9

I. Identify the reacting groups in the substrate.
II. Draw out a blank in the required *anti*-periplanar form (as in frame 8, but remember to replace X and Y with the reacting groups identified in I).
III. Insert the other groups into the blank, ensuring that the correct configuration is shown.
IV. Follow the reaction through to products.

Suppose that one were faced with the following problem.

Predict the configuration of the alkene resulting from the E2 elimination of hydrogen chloride (catalysed by hydroxide ion) from [1(*R*),2(*S*)] 1-chloro-1,2-diphenylpropane by following the stereochemical course of the reaction through to products. Use proper stereochemical diagrams.

Tackle the problem by breaking it down into parts (there will be four).

? *I. Identify the departing groups and the two carbons which will become the alkene carbons.*

* *

10

The departing groups are H (from C2), and Cl; the two carbons are Cl and C2.

? *II. Construct a blank with the reacting groups in the required* anti-*periplanar disposition and shown in the plane of the paper.*

* *

Your blank should be:

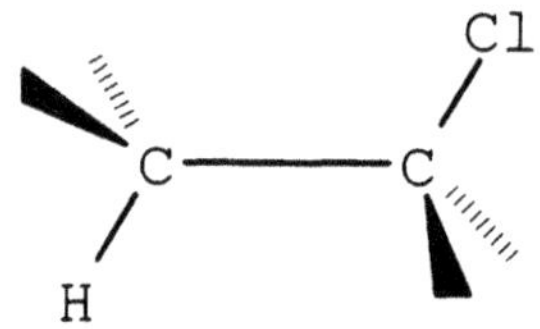

11 Note that there are three other orientations which you may have chosen for your blank, each of which would be equally correct/convenient:

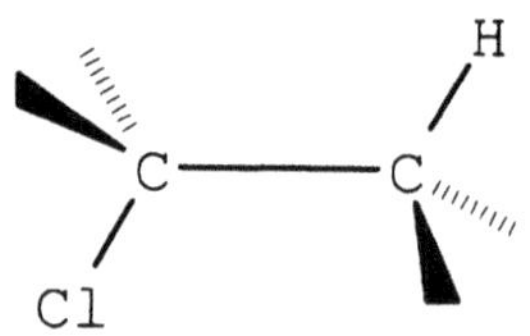

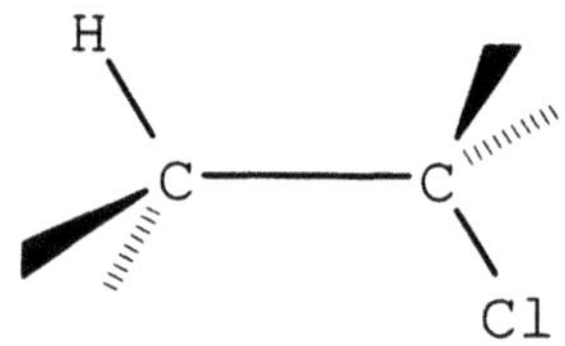

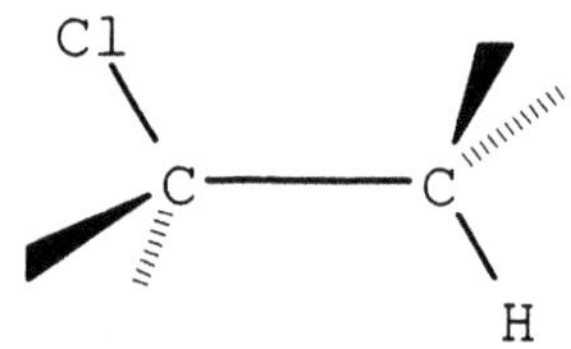

? *III. Starting with the first blank, fill in the unrepresented groups, carefully ensuring that you give each chiral centre the correct configuration as specified, to give the reacting conformation of the required stereoisomer of the substrate.*

* *

Your diagram of the reacting conformation should now look like:

12

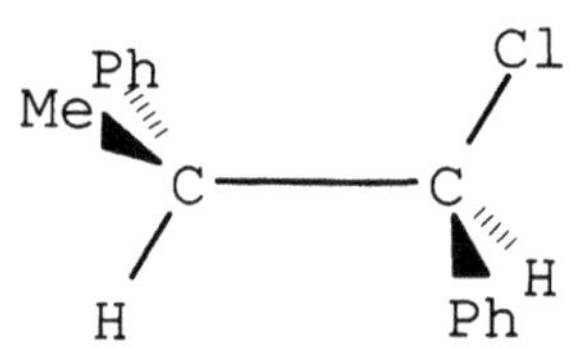

Priorities round C1:
–Cl 1; –Ph 2; –CHMePh 3; H 4

Priorities round C2:
–CHClPh 1; –Ph 2; –Me 3; H 4

? *IV. Now complete the process by following the reaction through to products.*

* *

Your reaction scheme should look like this, with the predicted product being the (E)-isomer:

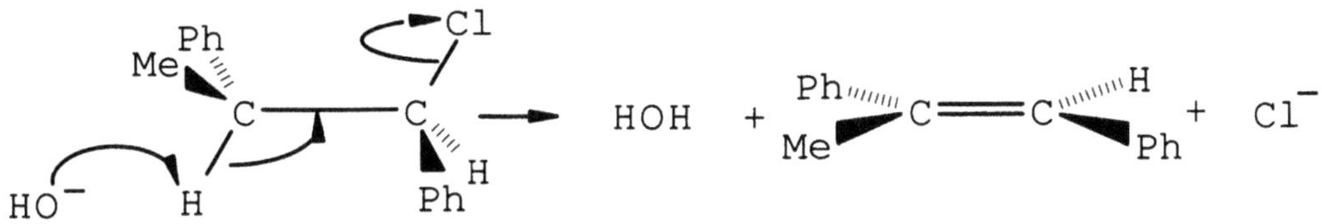

13 ? *Using the same method, follow through the base-catalysed E2 elimination of hydrogen chloride from the stereoisomer shown as a Fischer projection below. This time use the generalized symbol B for the base.*

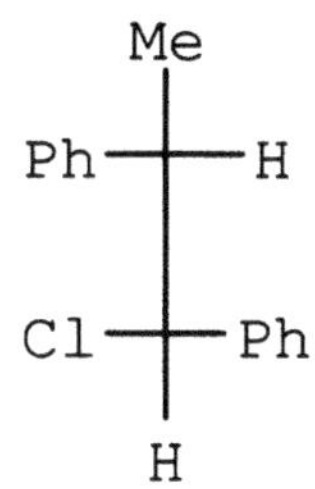

* *

The overall result should be as follows, and you should be able to identify from this scheme the correct result of separately carrying out the four steps outlined earlier.

14

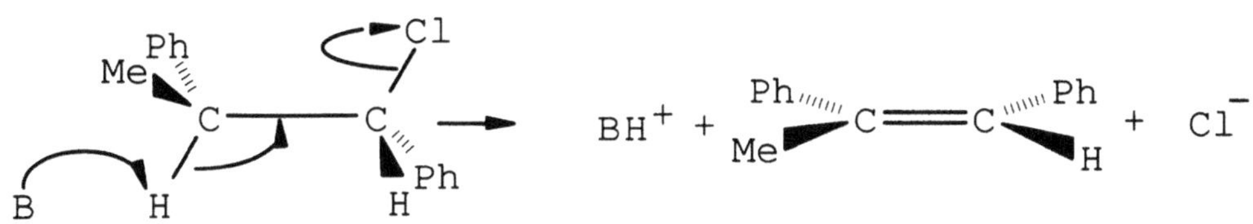

? *Is the stereoisomer of the substrate in this reaction the* threo- *or the* erythro-*form?*

* *

It is the erythro-*form.*

If you rotate the right-hand C atom in the reacting conformation above, about the C–C bond, then you can arrive at a conformation in which the two sets of identical groups (2 × H and 2 × Ph) are eclipsed. Thus this is the *erythro*-form (frame 4):

15

? *The* erythro-*form exists as a pair of enantiomers. What would be the stereochemistry of the alkene formed from the other enantiomer of the* erythro-*form?*

* *

The other enantiomer would also give the (Z)-form of the alkene.

16

If you consider the reacting conformations, then that for the one enantiomer must inevitably be the mirror image of that of the other. In both cases, therefore, the two phenyl groups will be on the same side of the 'plane of reaction', and the (*Z*)-isomer will be obtained.

? *What would be suitable substrates from which (by an E2 reaction) the (E)-alkene could be obtained?*

* *

Either enantiomer of the threo-*form.*

17

If you make a suitable diagram of the reacting conformation of each enantiomer of the *threo*-form, you will see that in each the phenyl groups are on opposite sides of the 'plane of reaction' and that each, therefore, leads to the (*E*)-alkene.

? *Make such a diagram of the [1(S),2(R)]* threo-*form.*

* *

Your diagram should look like:

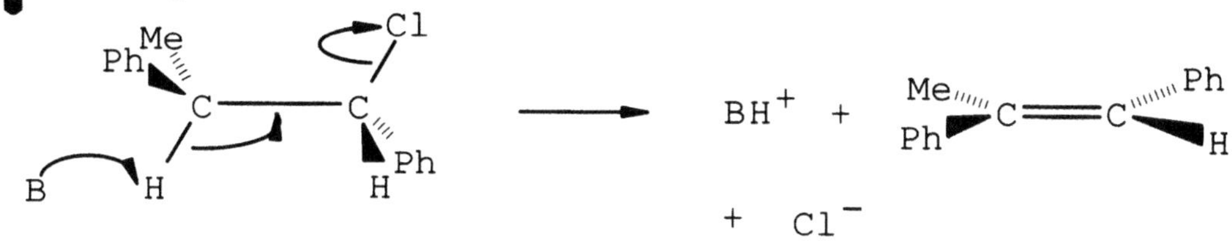

(The allocation of priorities about C1 and C2 of this reactant are given in frame 12.)

Consider now the stereochemistry of the E2 process in a rather different context.

18 The E2 elimination of HCl from menthyl chloride (I) slowly gives rise to a single alkene (2-menthene), whereas the corresponding elimination from neomenthyl chloride (II) is much faster and gives rise to two alkenes (2-menthene and 3-menthene).

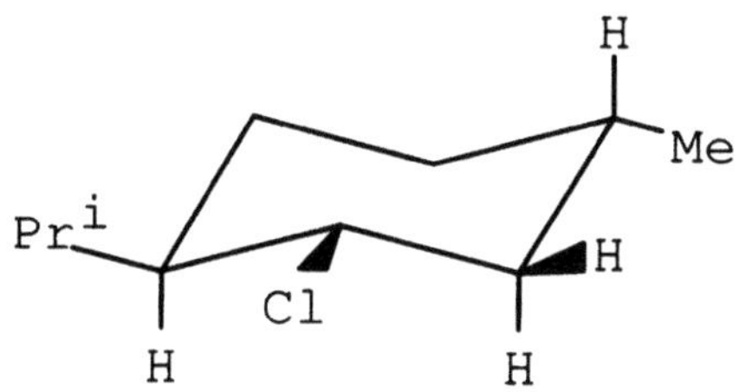

I

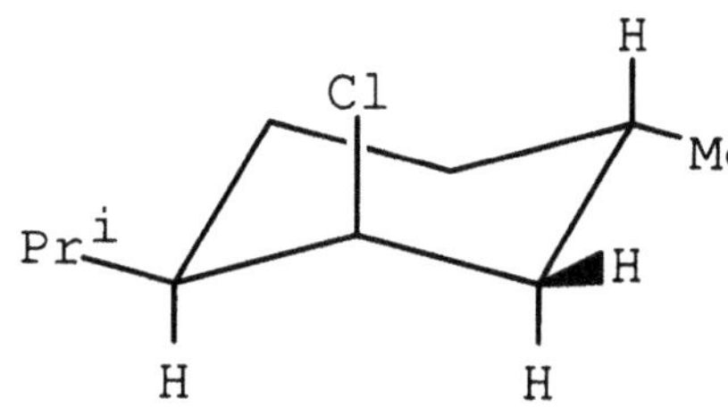

II

(Pr^i = isopropyl, $(CH_3)_2CH–$)

The ring numbering in menthyl and neomenthyl chloride is:

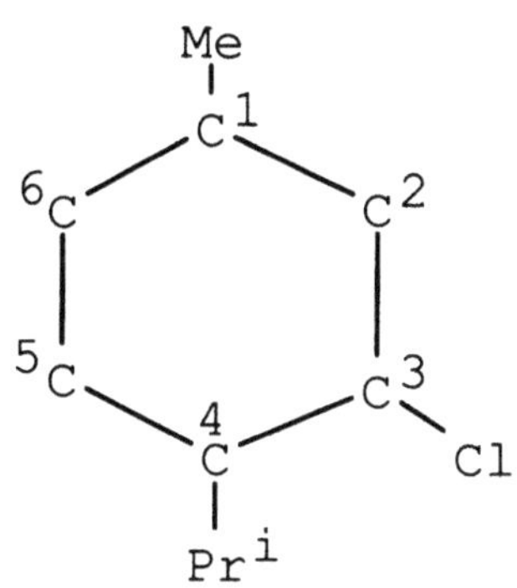

? *Account for the very slow elimination from menthyl chloride to give a single alkene.*

* *

Menthyl chloride is shown in its most stable conformation (I), with all of the groups equatorial. In that conformation there are no hydrogen atoms which are anti-*periplanar with the chlorine, so there can be no E2 elimination of hydrogen chloride.*

There will, however, be a low concentration of the less stable conformer (in equilibrium with the more stable conformer) in which all of the groups are axial. In this conformation – the only realistic alternative to the more stable one – there is ONE hydrogen which is anti-*periplanar with the chlorine, on position 2, and loss of this one, with the chlorine, leads exclusively to a double bond between carbons 2 and 3, i.e. the product is 2-menthene.*

The reaction will be slow, since there is only a very low concentration of the less stable conformer.

19

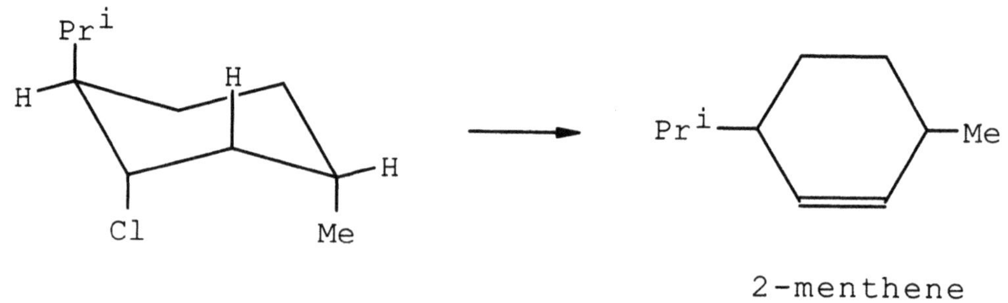

? *Account for the fact that the elimination from neomenthyl chloride is much faster, and leads to both 2- and 3-menthene.*

* *

The most stable conformer of neomenthyl chloride, in which both alkyl substituents are equatorial, is the one shown earlier (II). It has TWO hydrogens which are anti-*periplanar with the chlorine, one on position 2 and one on position 4. Thus either of these two can leave with the chlorine and both 2- and 3-menthene are produced.*

The reaction is not hindered by being restricted to a less stable conformer, present in low concentration, and so proceeds at a reasonable rate.

20

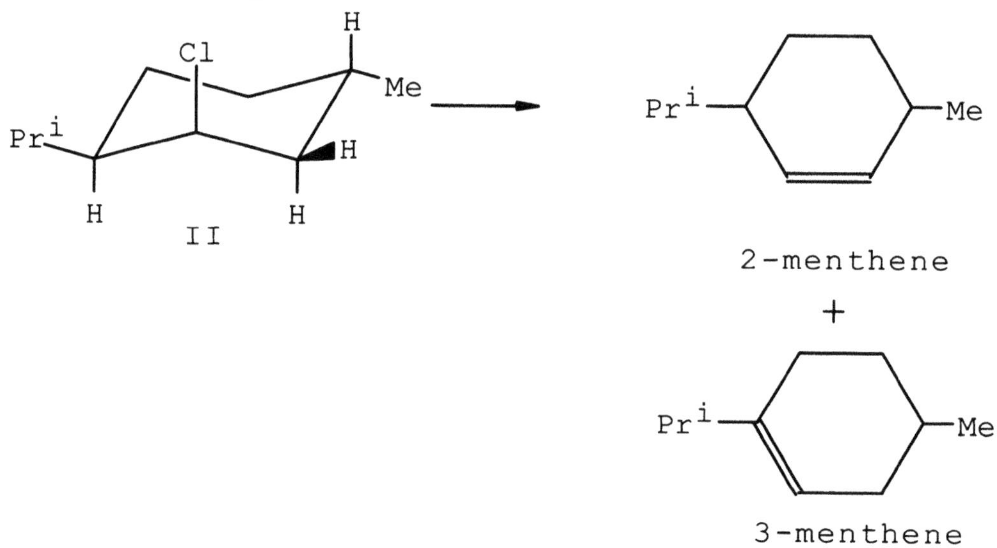

In frame 8 it was noted that in certain circumstances, where an *anti*-periplanar conformation could not be achieved, then an E2 elimination could occur via a *syn*-periplanar conformation. This is relatively rare, but the example which follows shows that it is possible.

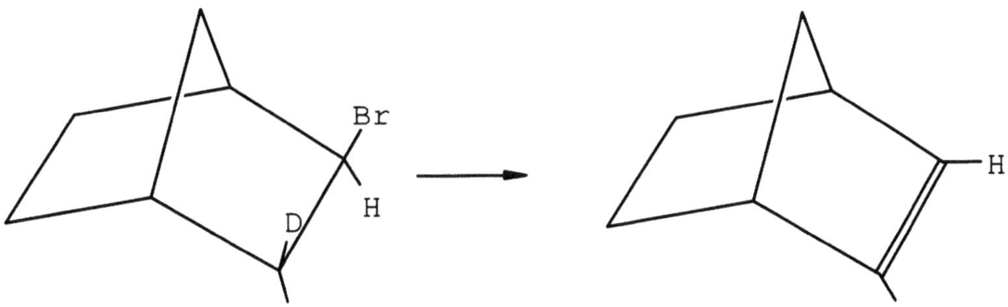

20 The bromine cannot depart with a hydrogen, since the adjacent C–H bond is far from being coplanar with the C–Br bond (dihedral angle *c.* 120°). However, the C–Br bond is coplanar with the adjacent C–D bond, albeit a *syn*-periplanar relationship, and reaction is possible.

Consider the two compounds (III) and (IV) below.

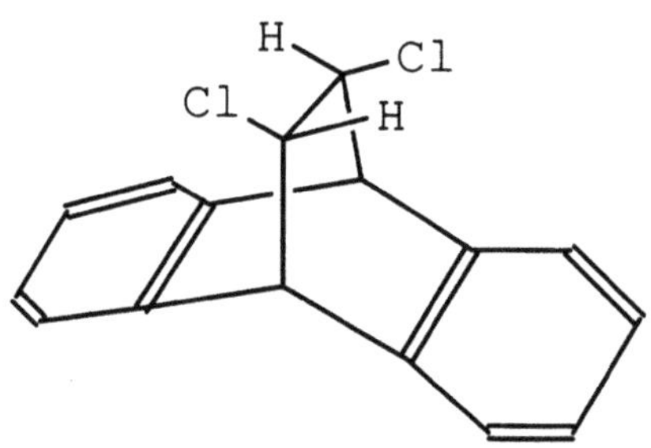

III

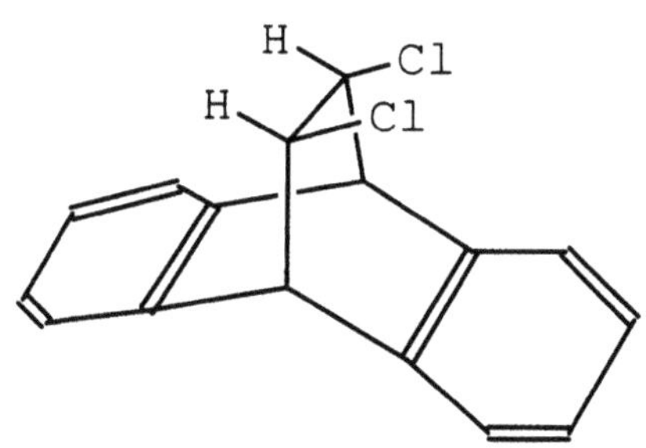

IV

? *What would be the product (if any) obtained by E2 elimination of HCl from compound (III)?*

* *

The product would be compound (V).

21

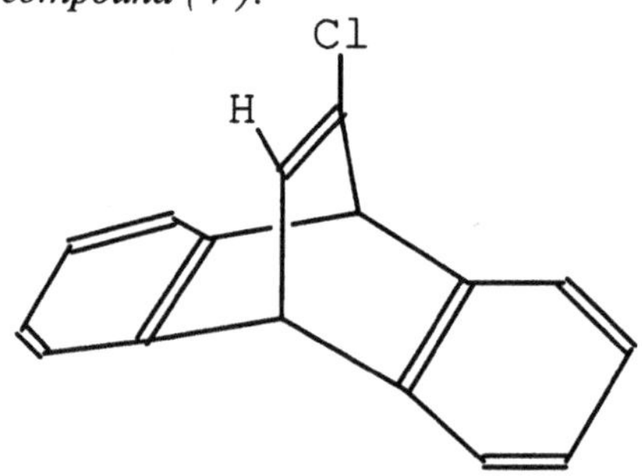

V

Although no *anti*-periplanar arrangement for the reacting H–C–C–Cl system is possible in III, the system is held permanently in a *syn*-periplanar arrangement, and reaction is thus possible.

? *What would be the product, if any, of E2 elimination of HCl from compound (IV)?*

* *

No E2 elimination is possible, since no coplanar arrangement, neither anti- *nor* syn-*periplanar, is possible for any H–C–C–Cl system.*

The only possible elimination route here would be an E1 route, which has no stereochemical constraints upon it. Thus there may, in practice, be a slow E1 elimination of HCl to give compound (V).

Finally, consider the case of **thermal eliminations**, which occur as a result of heating suitable substrates, rather than treating them with base (or acid) catalysts.

In these cases, the departing (leaving) groups act as an internal base, abstracting a proton from a position two carbons along the chain from the leaving group position.

22

For example, the ester butyl acetate (ethanoate) will eliminate acetic (ethanoic) acid to give 1-butene:

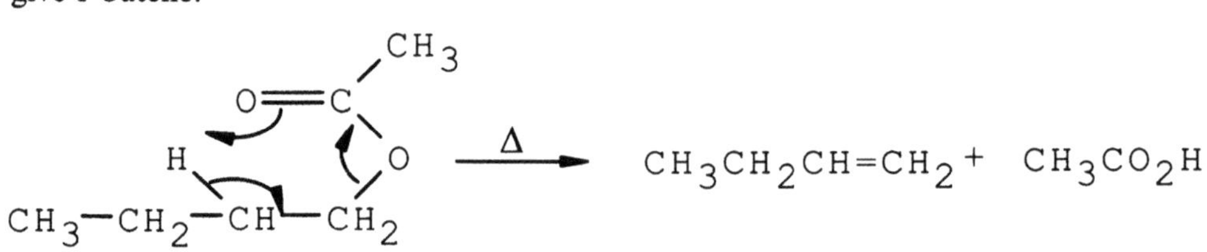

Although the transition state is essentially in the form of a six-membered ring, it will not (for several reasons) adopt the perfect chair conformation associated with cyclohexane. The part which is developing towards the alkene (H–C–C–O) will be close to an eclipsed conformation, reflecting the geometry of the alkene towards which it is progressing. Remember that the flat geometry of an alkene shows perfect eclipsing between the groups at either end of the C=C double bond. There is almost perfect staggering, of course, about all of the C–C bonds in the chair form of cyclohexane.

? *What would be the stereochemistry of the alkene (2-pentene) generated by the thermal elimination of acetic acid from (*S*)-2-pentyl acetate?*

* *

Both the (Z)- and the (E)-alkene will be produced, with the (E)-isomer as principal product.

A possible reacting conformation would be:

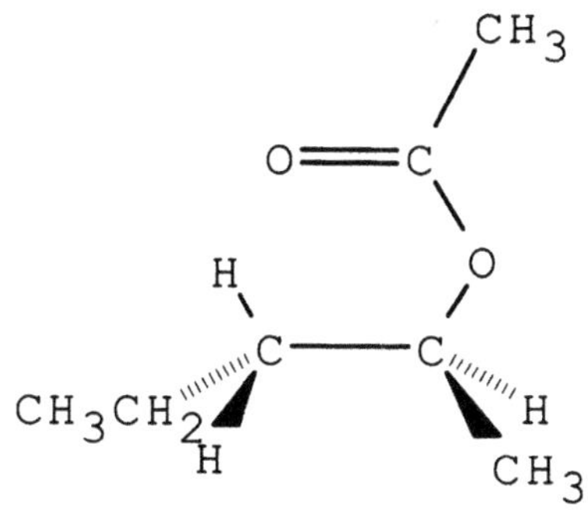

(Note how, once again, we have found it helpful to draw the diagram with the 'plane of reaction' in the plane of the paper.)

This would clearly lead to the (*E*)-alkene, as would perhaps be clearer if we drew a Newman projection of the system looking along the developing double bond:

23

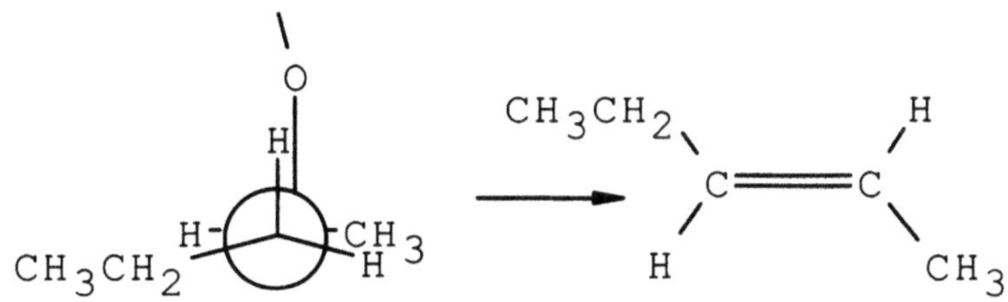

However, the acetate entity has a choice of abstracting either of the two hydrogen atoms on the CH_2 group to which the ethyl is attached. Had it chosen the other one, then the reacting conformation would have looked like:

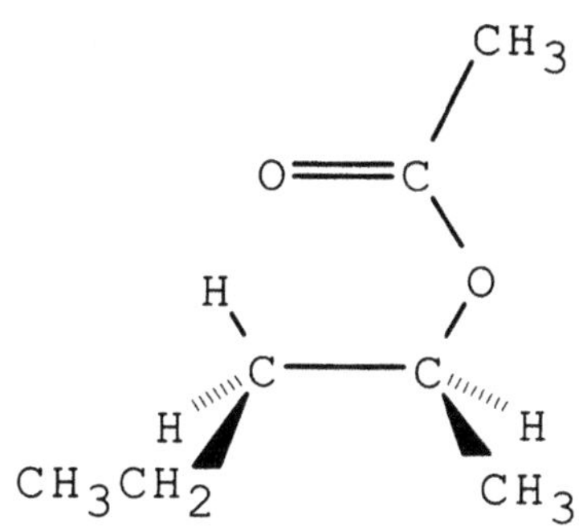

This would lead to the (*Z*)-alkene:

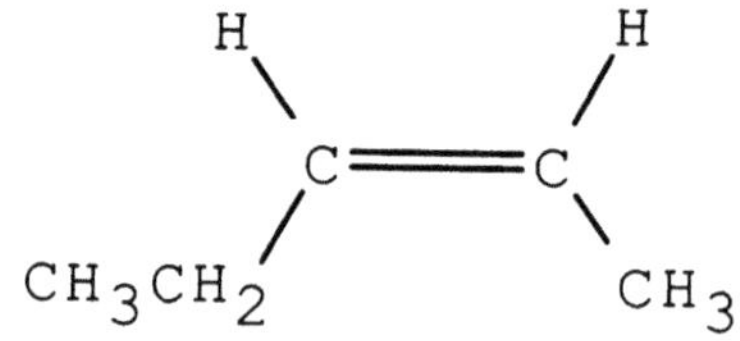

Both stereoisomers of the alkene can therefore be formed.

However, we can expect the (*E*)-isomer to be formed preferentially since the transition state leading to the (*Z*)-isomer has the two larger groups (ethyl and methyl) eclipsing one another:

23

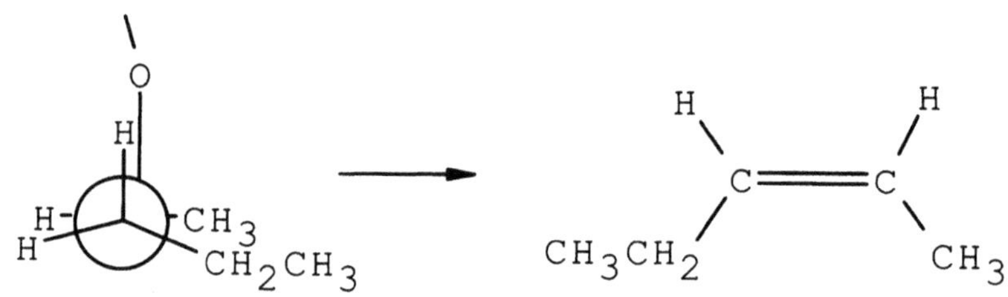

In contrast, that leading to the (*E*)-isomer has the two larger groups further apart (Newman projection, earlier in this frame). The transition state leading to the (*E*)-stereoisomer is therefore more stable than that leading to the (*Z*)-stereoisomer, and the former will form more rapidly.

? *Which stereoisomer of the resulting alkene (2-methyl-1-phenylbutene) would you expect to obtain from the thermal elimination of acetic acid ($MeCO_2H$) from [1-(*R*),2-(*R*)] 2-methyl-1-phenylbutyl acetate?*

* *

This time the acetate has only one possible hydrogen atom to take with it when it goes, and the product can only be the (E)-alkene.

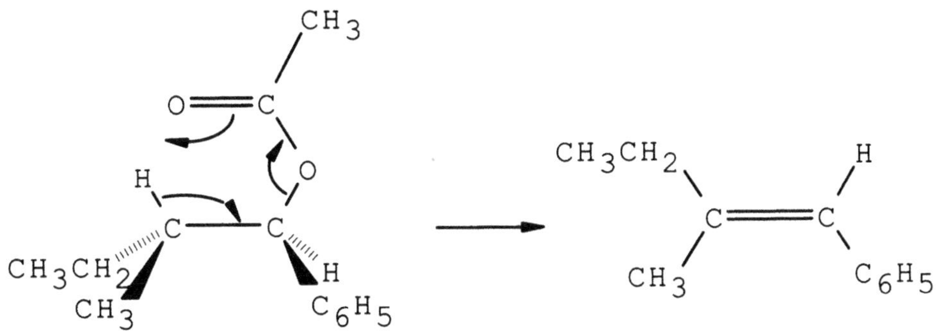

24 All of the elimination reactions which we have considered so far, E2 and thermal eliminations, have one feature in common; the departing groups all depart at the same time.

In E1 reactions, and in E1cB reactions, the groups depart one after the other. Once one has gone, the other has no 'recollection' of its stereochemical orientation, so there can no longer be any requirement for any particular stereochemical or conformational relationship between the departing groups.

Any choice between, say, two possible geometrical isomers of a product will depend solely on the relative stabilities of the transition states leading from the carbocation (E1), or the carbanion (E1cB), to the product alkene. The stabilities of the products are usually a good guide to these transition states, and one usually finds that the favoured product is the one in which large groups are *trans* to one another, i.e. as far away from each other as possible.

? *Predict the principal product of the E1 elimination of HCl from 1-chloro-1-phenylpropane, $CH_3CH_2CH(Cl)Ph$ (either enantiomer).*

* *

The principal product would be (E)-1-phenylpropene:

25

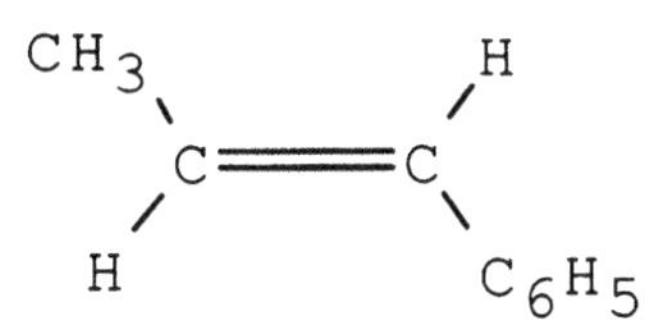

* *

This is the end of Programme 6, Dynamic Stereochemistry, Part II. If there are any difficulties which you cannot resolve from the programme itself, then seek advice from a tutor.

Programme 7

Thermodynamic and kinetic control

Understanding competing reactions in organic chemistry

There are many situations in organic chemistry where the treatment of one compound with a particular reagent may lead to two – sometimes several – products, each formed by a different reaction pathway (i.e. two or more reactions may take place at the same time, in competition with each other).

This can be generalized as follows, where A represents a reactant (or set of reactants) and B and C each represent a product (or set of products).

1

An example of such a situation would be the electrophilic addition of hydrogen chloride to propene where, in principle, two products are possible, 1-chloropropane or 2-chloropropane:

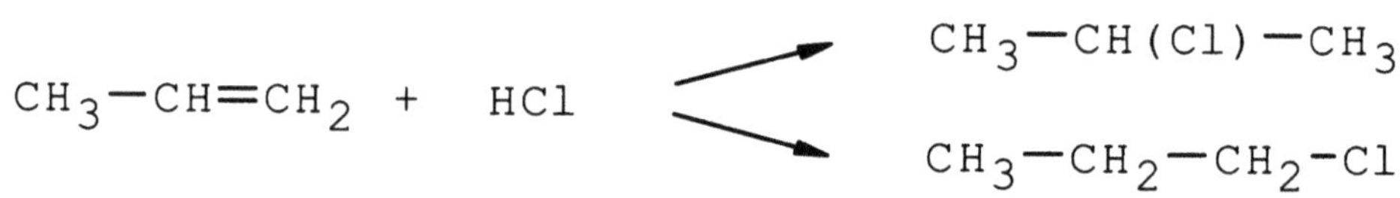

? *The nitration of nitrobenzene could, in principle, lead to three different products. What are they?*

* *

The three possible products are:

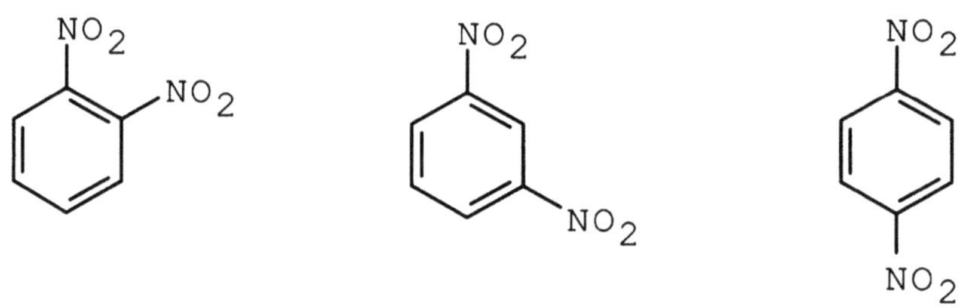

2

Clearly, organic chemists would like to be able to understand how the distribution of products is determined in situations like these, and, in new situations, to be able to predict what that distribution is likely to be.

Returning to the generalized representation of frame 1, there are TWO different outcomes which we could identify.

Firstly we could observe that either B or C is formed in the larger quantity, with only a smaller amount – perhaps none – of the other.

? *What would be the second possible outcome?*

* *

Both B and C could be formed in equivalent amounts, with no observable preference for one or the other.

Assuming that B and C differ in structure, as do the various products in the two examples considered earlier, then they form as a result of reaction taking place at different positions – or regions – in the reactant molecule (or **substrate**).

A reaction leading to the first outcome, with the formation largely of (say) B,
3 accompanied by smaller amounts of C, would be described as **regioselective**. **Regioselectivity** would be complete if no C were formed at all; high if only a little C had been formed; low if almost as much C as B had been formed.

If equivalent amounts of B and C form, then the reaction would be **non-regioselective**.

Sometimes the two different products, B and C in the generalized reaction scheme, have the same structure but are stereoisomers. In this case, then slightly different terms apply to reactions which lead to the different possible outcomes.

? *What would be the outcome of such a reaction described as* ***stereoselective****?*

* *

One of the possible stereoisomeric products (say B) would be obtained in larger amounts
4 *than the other (C).*

? *What would be the outcome of a reaction described as* ***non-stereoselective****?*

* *

5 *The stereoisomers would be formed in equivalent amounts.*

Thermodynamic control

Referring still to the generalized reaction, consider what would happen if the reaction leading from A to B, and the reaction leading from A to C, were both reversible:

6 Clearly, equilibria would be set up for both reactions, and, in effect, equilibrium would be established between C and B.

? *Denoting the equilibrium constant between A and B as* K_{ab}*, and that between A and C as* K_{ac}*, derive an expression for the equilibrium constant between B and C,* K_{bc}*. Assume, for simplicity, that A, B and C each represent one mole of a single compound, rather than a set of reactants or products.*

* *

$K_{bc} = K_{ac}/K_{ab}$

This expression can be derived as follows:

$$K_{bc} = [C]/[B]$$

$$K_{ac} = [C]/[A]$$

$$K_{ab} = [B]/[A]$$

7 $\therefore \quad K_{ac}/K_{ab} = [C][A]/[A][B] = [C]/[B] = K_{bc}$

It is clear, therefore, that the equilibrium constant K_{bc} will determine the relative quantities of B and C when reaction is complete. In other words, the selectivity is determined by K_{bc} .

Equally, since K_{bc} is a function of K_{ac} and K_{ab} , one could say that the selectivity is determined by K_{ac} and K_{ab} together (i.e. by the ratio K_{ac} / K_{ab}).

What (in terms of the equilibrium constants involved) will be the requirements for the reaction to be non-selective (again, for simplicity, assume that A, B and C are single compounds, each symbol representing one mole)?

* *

K_{bc} *will have to take the value of 1 (i.e.* $K_{ac} = K_{ab}$*).*

8 For values of K_{bc} close to 1, the reaction will be of low selectivity.

What will be the requirements for the reaction to be of high selectivity?

* *

K_{bc} *will need to be well removed from the value of 1 – substantially larger or substantially*
9 *smaller than 1 (i.e. either* K_{ac} *is much larger than* K_{ab} *or* K_{ab} *is much larger than* K_{ac}*).*

If K_{bc} is larger than 1, then the reaction will be selective in favour of C.

Why is the reaction still regarded as selective if the equilibrium constant, K_{bc}*, is less than 1?*

* *

Because the reaction will then favour the other product; it will be selective in favour of B.

Reactions which are regio- or stereoselective because they are reversible, and the outcome of competition between reactions is determined by equilibrium constants, are said to be under **thermodynamic control**.

10 Note that for thermodynamic control to operate, the reactions must be reversible; the rates of the 'forward' reactions must be commensurate with those of the 'backwards' reactions. If this condition is not met, then equilibrium cannot be established (within a reasonable time limit) and the relative proportions of various products cannot be determined by equilibrium constants.

Occasionally reactions are met which are under thermodynamic control only at higher temperatures; only at these higher temperatures can equilibrium be reached quickly enough (i.e. within a reasonable time limit) for the amounts of the various products to be determined by the equilibrium constants. An example of a reaction like this will be met later in this programme (frame 13).

Kinetic control

11 Refer back again to the generalized reaction in frame 1, and consider the question of the relative quantities of B and C which will form if the reactions are *not* reversible. In other words, consider what the effect on product distribution would be if the forward reactions proceeded at a reasonable rate, but the backwards reactions were too slow to be observed – effectively of zero rate.

? *What would determine the relative quantities of B and C which were formed under these circumstances where, in effect, the reverse reactions did not occur?*

* *

The product which formed the faster would be the one which would form to the greater extent.

There would, in effect, be a 'race' between the competing reactions. In other words, the relative rates of product formation would determine the product distribution, and these reactions are therefore said to be under **kinetic control**.

12 Consider the sulphonation of naphthalene, where a hydrogen atom on the naphthalene ring system is replaced by a sulphonic acid group.

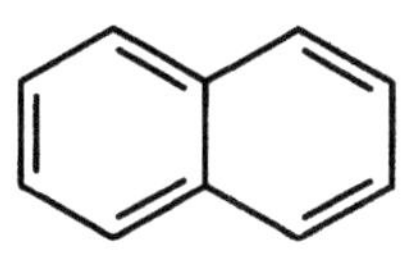

naphthalene

$$C_{10}H_8 + H_2SO_4 \rightarrow C_{10}H_7SO_3H + H_2O$$

? *How many different products would be possible in this reaction?*

* *

Two products (and only two) may be formed here, naphthalene-1-sulphonic acid and naphthalene-2-sulphonic acid:

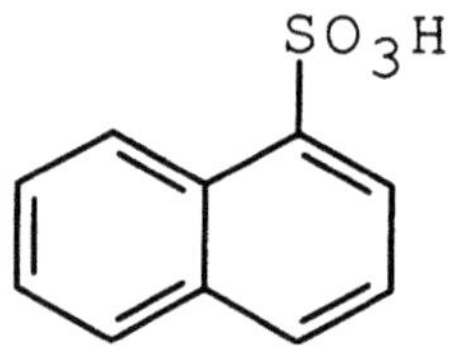

and

naphthalene-1-sulphonic acid

13

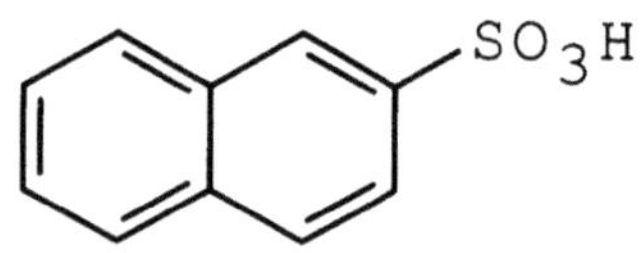

naphthalene-2-sulphonic acid

If this reaction is carried out at 80°, it is regioselective and the principal product is naphthalene-1-sulphonic acid. However, if the reaction is carried out at 160°, it is still regioselective but the principal product is now naphthalene-2-sulphonic acid.

If naphthalene-1-sulphonic acid is simply heated to 160°, under the same conditions as the sulphonation reaction, then it is largely converted into naphthalene-2-sulphonic acid.

? *Offer an explanation of these data in terms of the ideas of thermodynamic and kinetic control, introduced above.*

If this seems difficult at first, then refer to frame 10, paying particular attention to the final paragraph.

* *

This is a case where reaction rates are too slow to permit equilibrium to be achieved (within a reasonable time limit) at the lower temperature (80°). Consequently, the reaction is under kinetic control at that temperature. However, as the temperature is raised to 160°, reaction rates (in both directions) increase and equilibrium is rapidly achieved. We therefore see the result of thermodynamic control.

14 This explanation is supported by the result of heating the naphthalene-1-sulphonic acid; the interconversion of the two isomers becomes reversible at the higher temperature, and equilibrium is achieved within a reasonable time limit.

It is important to note that the results of thermodynamic and kinetic control are not the same in this case. Different products are obtained for each.

? *Which of the two products is the more stable?*

* *

Naphthalene-2-sulphonic acid is the more stable product, since it is the principal substance present at equilibrium.

15 It therefore follows that naphthalene-1-sulphonic acid is the less stable product. It forms more rapidly at both temperatures, but at 80° the reaction cannot reverse rapidly enough to achieve equilibrium within a reasonable time, whereas at 160° it can.

? *Does it follow that, in all sets of competitive reactions, the less (or least) stable product will form more rapidly?*

* *

No. There is no reason at all why this should be so.

There are, in fact, many cases where the more stable product is also the one which is formed the faster, i.e. where the result of kinetic control is essentially the same as the result of thermodynamic control.

Thus, the elimination of hydrogen bromide from 2-bromobutane gives largely 2-butene:

16

$$CH_3CH_2CH(Br)CH_3 \xrightarrow{NaOEt/EtOH} \underset{80\%}{CH_3CH{=}CHCH_3} + \underset{20\%}{CH_3CH_2CH{=}CH_2}$$

This reaction is under kinetic control, and regioselective in favour of 2-butene. The more stable alkene is the one which forms faster, so a similar result would have been observed had the reaction been under thermodynamic control.

Free energy profiles (not to scale) for these two competing reactions are illustrated below:

16

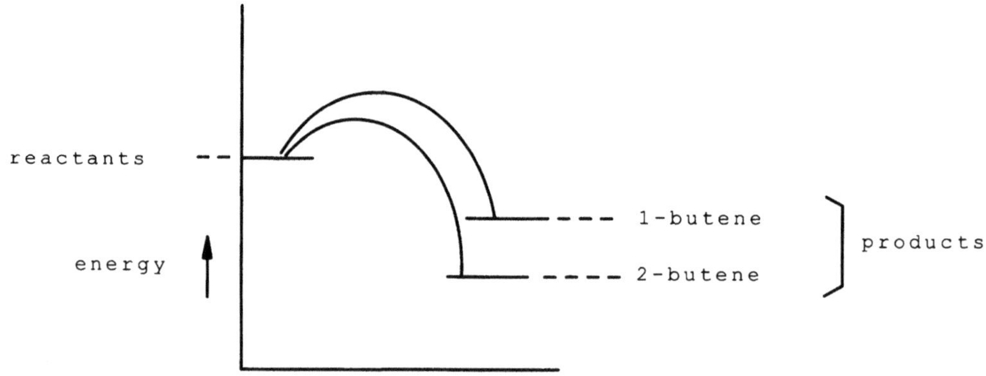

? *Sketch a similar set of profiles for a situation where reactants may give rise to two products, with the less stable one forming the faster.*

* *

17

Your diagrams should look like:

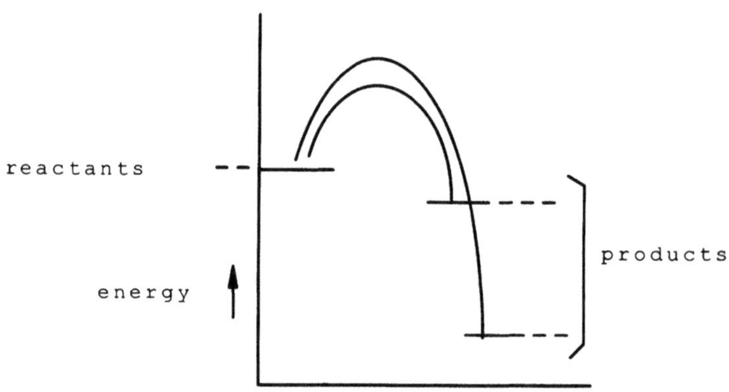

* *

Explaining and predicting (kinetic control)

Most competing organic reactions – but not all (e.g. naphthalene sulphonation at 160°) – are under kinetic control. Explanations of regio- or stereoselectivity must therefore be kinetic in nature, and must concern themselves with the relative rates of the competing reactions. Since differences in energy of activation (E_{act}) largely determine rate differences in these competitive situations, it is with E_{act} that such explanations must be concerned.

18 ? *On any one of the free energy profile curves earlier (i.e. any one pathway from the diagrams of frames 16 or 17), identify the difference in standard free energy ($\Delta G°$), which will determine the equilibrium constant, and the free energy of activation (for simplicity call this E_{act}) which will determine rate.*

* *

You should have identified these quantities as follows:

19

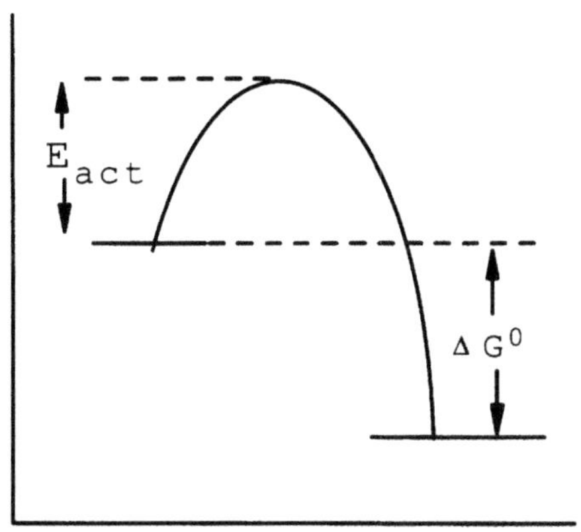

Note that these are two quite different quantities.

Note also that the 'hump' on the curve, at the top of which the transition state 'sits', is sometimes referred to as the activation barrier.

$\Delta G°$ is the difference in standard free energy between the ground state (reactants) and the products.

? E_{act} *may also be described as the difference between the energies of two 'states'. What are these two states?*

* *

The ground state (reactants) and the transition state.

The transition state – a hypothetical, transient 'species' – is the combination of reactants corresponding to the top of the profile curve, and the reactants must acquire at least this energy in order to 'pass' over the top of the curve and become products.

20

Note that when we are considering a set of competing reactions involving the same reactants, the ground state is the same for all pathways. In order to compare the activation energies for the two (or more) pathways, therefore, since the ground state is common to all, we need only consider the transition states. If one transition state is of higher energy – i.e. is less stable – than the other(s), then reaction will proceed more slowly via that transition state – and that pathway – than by the other(s). Conversely, the reaction will proceed most rapidly via the pathway(s) with the most stable – lower energy – transition state(s).

? *How, then, for a reaction under kinetic control and with two competing pathways, could the preferred product be predicted? What would we need to know?*

* *

The preferred product would be that which is formed by the pathway with the lowest energy – most stable – transition state. That is, by the reaction pathway with the lowest activation barrier. We would need to know the relative stabilities of the two transition states.

21

The prediction of the relative stabilities of transition states is not always easy. Remember that a transition state has no stable existence; we cannot isolate it and directly determine its precise structure and energy. We usually have to make sensible 'guesses' about structures, therefore, and then about relative energies.

Consider again the elimination of HBr from 2-bromobutane (frame 16). This would be a single step (E2 – Programme 4, frame 12) process.

? *Knowing what the reactants and products are, for this reaction, can you say anything at all about the nature of the transition state for each of the two pathways?*

* *

They have to be intermediate in character between reactants and products, with the double bond developing in the appropriate place.

Thus, we could represent the transition state leading to 2-butene as:

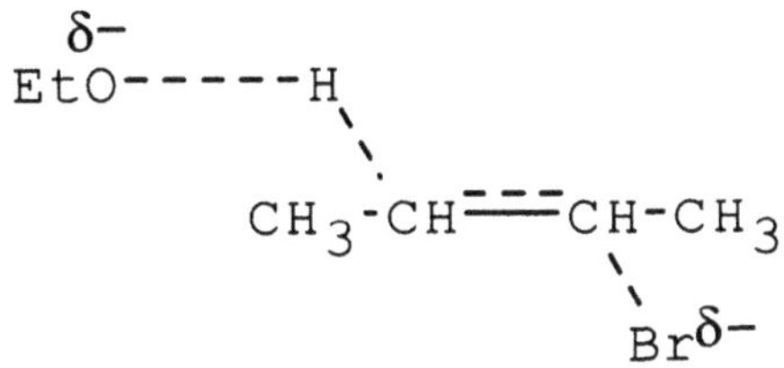

22 Developing (or stretching) bonds are represented as broken lines; no attempt has been made to represent the three-dimensional geometry of the species.

The transition state leading to 1-butene would be:

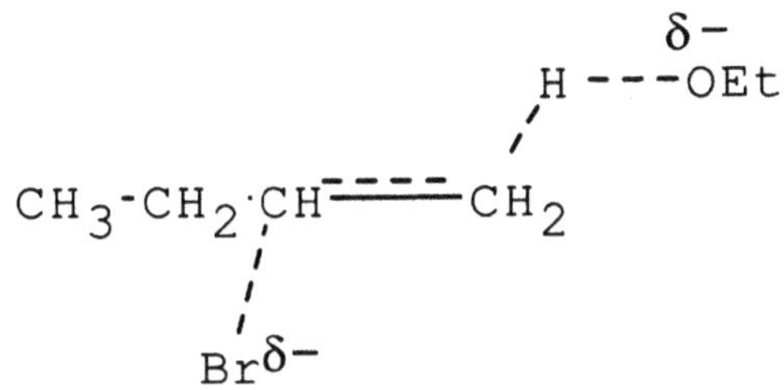

? *How many alkyl groups are attached to the carbon atoms at either end of the developing double bond in each of these transition states?*

* *

Two (both methyl) in the case of the transition state leading to 2-butene; only one (ethyl) in the case of that leading to 1-butene.

We know that alkenes with alkyl groups attached to the carbon atoms at either end of the double bond are of lower energy than those without (or with fewer). This is why 2-butene (with two alkyl groups attached to doubly bonded carbons) is more stable than 1-butene (with only one such alkyl group).

23 It is therefore reasonable to suppose that if the carbon atoms attached to the developing double bonds in a transition state are also attached to alkyl groups, then these alkyl groups will similarly stabilize that transition state.

Thus we can expect the transition state leading to 2-butene to be more stable than that leading to 1-butene, and the reaction to be regioselective in preferentially forming 2-butene by the faster pathway.

Sometimes the situation is rather more complex than those which we have so far looked at in detail, because the competing pathways to different products may involve several steps. A case in point would be aromatic substitution.

? *Write down, in outline, the steps involved in the electrophilic substitution of benzene. Use sulphonation as your example, with sulphur trioxide as the electrophile.*

* *

You should have written down two steps as follows:

24

The activation energy for a several step process is determined by the transition state of the rate determining (slowest) step (compared with the ground state for the first step), so to make rate predictions, we have to know which is the rate determining step. *In this case it is the first step.*

? *Write down similar mechanisms for the sulphonation of naphthalene to give both products (frames 12 and 13).*

* *

You should have written down mechanisms as follows:

25

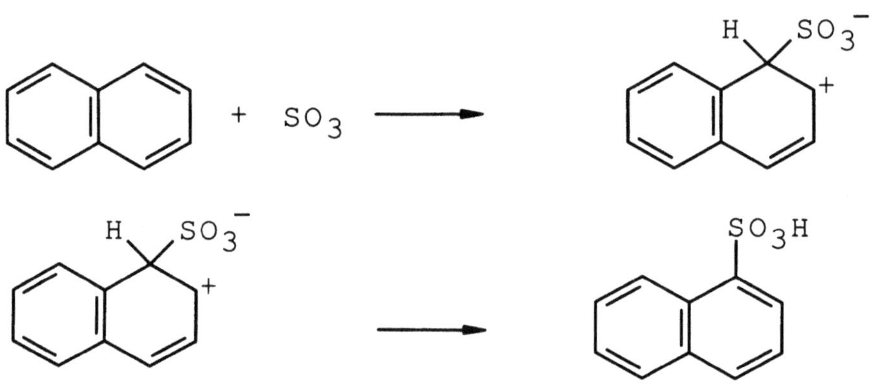

and:

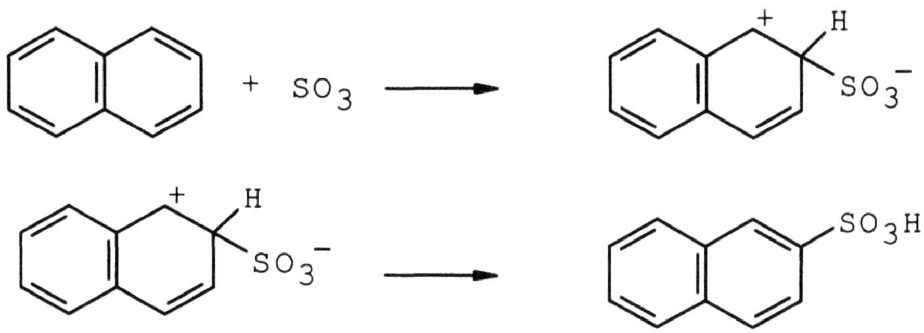

25 Since the first step in these processes is rate determining, we use the product of that step – the reactive carbocation intermediate – as a guide to the nature of the transition state, just as we did in the case of the E2 elimination of HBr from 2-bromobutane (frame 23).

The transition state will have much of the character of the carbocation, and whatever might stabilize the cation will stabilize the transition state which it resembles. Thus, if there is any factor which will make one of the cations more stable than the other, then that factor will have a similar effect upon the transition states.

? *Write down all the resonance forms (Programmes 1 and 2) that you can for the two carbocations, and then compare the two cases.*

* *

The resonance forms for each cation are as follows.

Leading to naphthalene-1-sulphonic acid:

26

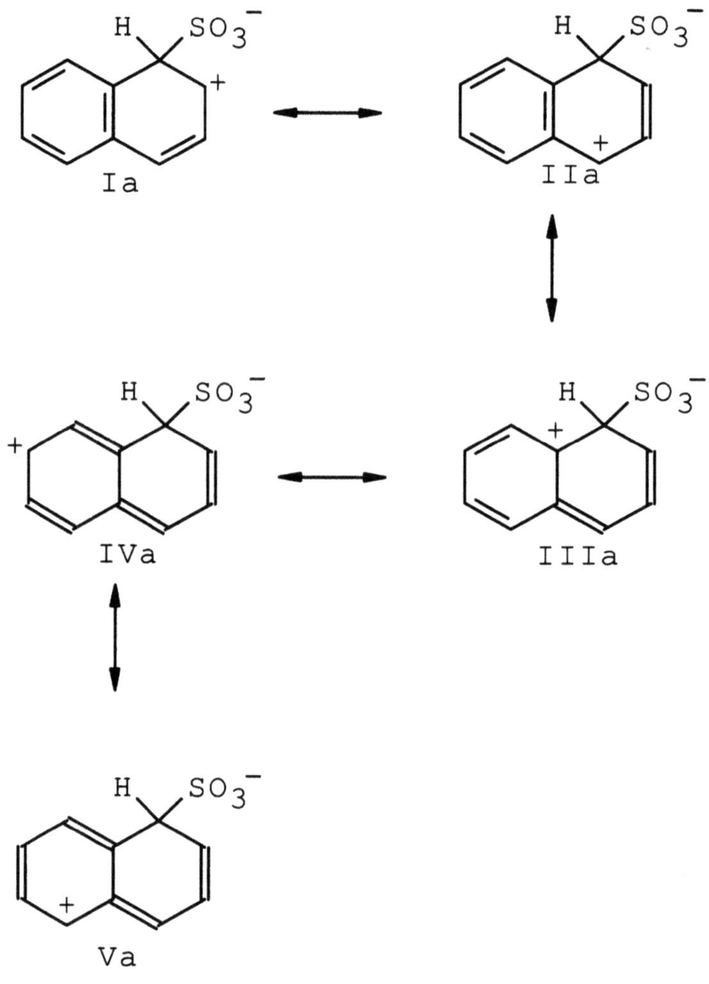

Leading to naphthalene-2-sulphonic acid:

26

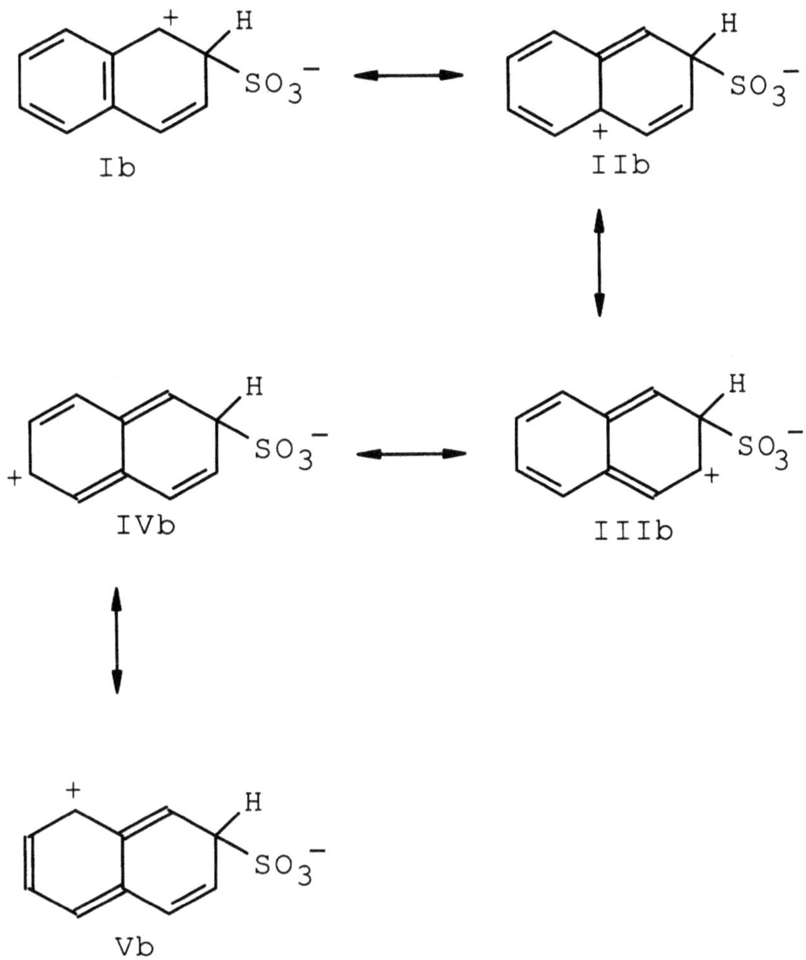

The next step is to decide which of these two cations is the more stable. In the case of cations derived from polycyclic aromatic systems, one can pick as the more stable cations those which contain 'complete' benzene systems. The more of these there are, the more stable will be the cation. The transition state leading to the more stable cation will also be the more stable one, and will represent the pathway of lowest activation energy, and therefore the one which proceeds most rapidly and leads to the favoured product.

? *Which of the two cations will be the more stable?*

* *

27

That leading to naphthalene-1-sulphonic acid will be the more stable since it it has two resonance forms (Ia and IIa) with 'complete' benzene rings, whereas that leading to naphthalene-2-sulphonic acid has only one (Ib).

? *Which would be the preferred product in this reaction, then, if it is kinetically controlled?*

* *

28

The preferred product would be naphthalene-1-sulphonic acid, since this would form via the more stable of the two transition states, the relative stabilities of which are indicated by those of the carbocations they lead to in the rate determining step.

Note that this matches what actually happens at 80° (frame 13), but not what happens at 160° when thermodynamic control takes over.

The purpose of looking in detail at these examples of reactions under kinetic control is to illustrate how predictions can be made as to their regio- (or indeed stereo-) selectivity.

Provided that one is satisfied that an organic reaction is under kinetic control, then one must estimate the relative stabilities of the transition states of the rate determining steps in the various competing reactions which lead to the different products. To do this one uses the product of that step as a guide. The preferred pathway is then that one (or ones) with the lowest energy transition state(s).

END

9 780412 558306

GPSR Compliance

The European Union's (EU) General Product Safety Regulation (GPSR) is a set of rules that requires consumer products to be safe and our obligations to ensure this.

If you have any concerns about our products, you can contact us on ProductSafety@springernature.com

In case Publisher is established outside the EU, the EU authorized representative is:

Springer Nature Customer Service Center GmbH
Europaplatz 3
69115 Heidelberg, Germany

Batch number: 10607930

Printed by Printforce, the Netherlands